QUANTITATIVE TECHNIQUES
IN GEOGRAPHY:

AN INTRODUCTION

Quantitative Techniques in Geography:

An Introduction

BY

ROBERT HAMMOND

AND

PATRICK McCULLAGH

SECOND EDITION

CLARENDON PRESS · OXFORD

Oxford University Press, Walton Street, Oxford OX2 6DP

London Glasgow New York Toronto
Delhi Bombay Calcutta Madras Karachi
Kuala Lumpur Singapore Hong Kong Tokyo
Nairobi Dar es Salaam Cape Town
Melbourne Auckland

and associates in
Beirut Berlin Ibadan Mexico City Nicosia

CASE BOUND   ISBN 0 19 874066 2
PAPERBACK   ISBN 0 19 874067 0

© *Oxford University Press 1978*

First edition 1974
Second edition 1978
Reprinted 1980, 1982

Set by Hope Services, Wantage,
and Printed in Great Britain by
Richard Clay (The Chaucer Press) Ltd, Bungay, Suffolk

"The quantification of theory, the use of mathematics to express relationships, can be supported on two main grounds. First, it is more rigorous. Second, and more important, is is a considerable aid in the avoidance of self-deception."

Ian Burton, *The Canadian Geographer* (1963).

PREFACE TO THE FIRST EDITION

THE writers of any textbook are faced with problems of selection and organization. This book can outline only a sample of the available elementary numerical techniques. The field is very wide indeed and the choice has necessarily been arbitrary. Nevertheless, throughout the book, an attempt has been made to draw examples from as many aspects of human and physical geography as possible. It is hoped that variety in presentation, and breadth of coverage, will be sufficient to provide insight for the student into the many ways whereby quantitative techniques may be useful in the presentation and solution of geographical problems.

Explanation of the derivation of formulae has not generally been attempted, and this has influenced the sequence of chapters. The mathematical purist would, no doubt, want to deal with regression before correlation; but in this book it is appropriate to present correlation coefficients (which are used in least squares regression formulae) before regression (which is not used in the correlation formulae). Notation has been deliberately simplified to render formulae less formidable in appearance—e.g. by the omission of summation indices.

There is no simple 'right' way of dividing the subject into chapters. The sequence in this book is based on the different kinds of problem with which the geographer is faced—e.g. the summary of data relating to a single place (Chapter 1) is a different problem from that of summarizing spatial distributions (Chapter 2), although both may use analogous techniques, such as the Lorenz Curve, which therefore appear in two chapters. To that extent, the book is 'problem-oriented' rather than 'techniques-oriented', although, as the title indicates, it is concerned with the exposition of techniques, and individual problems are made to 'fit' the techniques.

We especially wish to thank Mr D. Kaye of the Extra Mural Department of the University of Manchester for his help and advice on some of the inferential techniques included.

We are also indebted to the Literary Executor of the late Sir Ronald A. Fisher, F.R.S., to Dr Frank Yates, F.R.S., and to Longman Group Ltd., London, for permission to reprint tables from their book *Statistical Tables for Biological, Agricultural and Medical Research*. We are also grateful for permission to reproduce tables and figures from other sources: details are given in the underlines and bibliography.

R.H.
P.S. McC.

PREFACE TO THE SECOND EDITION

THIS edition has been enlarged and revised as the result of four years' experience, and in order to include techniques developed since 1974, such as the Linear Nearest Neighbour Index. We have also added a number of short sections on topics in which student understanding is often wrongly assumed: examples are the use of surrogates and the meaning of statistical independence. We have extended the text in several places where we now feel the first edition did not take the student far enough; for example, in the transformation of data, and in calculations based on exponential models related to time series. We have also introduced, by simple graphical methods, the principles of linear programming, to provide a foundation of understanding from which students may move on to more sophisticated applications, using library computer programmes. We are aware of the constraints and practical limitations in the use of some of the techniques, and have endeavoured to face up to these problems—for instance, by a discussion of boundary definition in calculating the Nearest Neighbour Index.

Finally, we wish to acknowledge the advice of colleagues in the Mathematics department of Trent Polytechnic, which has been of considerable value in both the revision of the original text, and in the writing of additional material.

R.H.
P.S.McC.

CONTENTS

INTRODUCTION

I.   THE AIM

RECENT years have seen a marked increase in the application of models and of many different descriptive and inferential statistical techniques to geographical problems. Some of these techniques are exceedingly complex. Others are relatively simple and may be applied to fieldwork normally carried out in either physical or human geography by any student. The object of this book is primarily to give a working understanding of the simpler quantitative techniques *to those who have no background of mathematics.*

The contents of this book are the result of experience in teaching students how to apply analytical techniques to data acquired directly through work in the field, from published material of various kinds (e.g. Census reports), and the analysis of maps. It has been found that, provided the only assumptions made are an ability to add, subtract, multiply, and divide, students are able to understand the uses and limitations of elementary statistical methods and apply them correctly. The conclusions are often salutary for the geographer. From experience it has often been found that an apparently meaningful pattern (e.g. of land use or farm economy), when tested statistically, could have occurred by chance more than 5 times in 100. Statistical analysis therefore develops in the geographer a much more critical approach. He is forced to think more rigorously and precisely, and is less inclined to make vague generalizations based on insufficient evidence.

'Quantitative geography' is here to stay. The revolution is over. It is not a 'new' geography in the sense that it alters the nature of the discipline itself; but only in that it offers new techniques—for the solution of old problems as well as new ones—techniques which have brought a new rigour into geographical thinking, and have made possible the exploration of new fields. The acquisition of these techniques, or the understanding of them when used in published material, can be very difficult for those whose background contains no mathematics. An attempt is made in this book to explain in simple language some of the basic concepts and methods involved which are of special interest to the geographer. It is written as an introduction to the use of quantitative techniques, as an aid to the understanding of published literature, and as a basis from which to proceed to more advanced study.

The techniques outlined may usefully be divided into three categories:

1. *Descriptive techniques*, such as a coefficient of diversification, which only describes, and about which no question of probability arises.

2. *Inferential techniques*, where some measure is necessary to estimate the extent to which a sample may be regarded as representative of the whole, or the degree to which a collection of data supports a hypothesis.

3. *Model-making techniques*, which are playing an increasingly important part in quantitative geography, and which may be both descriptive and inferential.

II.   DESCRIPTIVE TECHNIQUES (CHAPTERS 1–3)

These are used to summarize information (large numbers of values) about places, areas, location patterns, or trends and fluctuations through time, in order to provide a single comprehensible index (or graph), and thereby facilitate accurate descriptions and comparisons. The arithmetic mean (often called the 'average') is a simple example of this technique.

III.   INFERENTIAL TECHNIQUES–INCLUDING SAMPLING (CHAPTERS 4–9)

Sampling is very important for the geographer as so much of his data must inevitably come in this form. (It is impossible to measure all the pebbles on the beach. Nor is it practicable to question everyone in an urban area, or visit every farm in a county.) A sound knowledge of sampling techniques and the implications of the use of samples is therefore essential.

Inferential statistics are involved whenever the question of probability arises. For example, they may be used to test the level of confidence with which a sample may be deemed to be representative of the whole; or the probability of two separate independent samples coming from the same population; or to test for randomness in spatial distributions. If a special pattern of land use is disclosed as a result of fieldwork, before conclusions are drawn it is desirable to be able to show statistically the extent to which the pattern may not be due to chance.

IV.   MODEL-MAKING TECHNIQUES (CHAPTER 10)

Some models are constructed to be as exact an image of some aspect of reality as possible. They are a kind of 'picture' in two- or three-dimensional form. These kinds of model have been termed 'iconic', from the Latin *iconicus*, meaning of the nature of a portrait. Examples, which have been used for many years, are the relief model of a drainage basin, or the rather more theoretical topographical map. More recently geographers have explored the use of more abstract kinds of model, frequently employing mathematical techniques, in their attempt to describe, explain, or predict interrelationships. But because reality is generally very complex, and because there may be many phenomena present, not all of which are judged relevant to a problem being studied, some element of selectivity is generally required to reduce the complexity sufficiently to enable a model to be constructed. A model of this kind has been defined as 'a simplified structuring of reality', an attempt to see the wood by eliminating some of the trees.

It follows that the way in which reality is simplified and structured implies a theory. It may be an *a priori* model because it has been constructed to *test* a theory, or, more often, it is the end product of a series of experiments, or a concept resulting from experience. A model may also be only descriptive. Some geomorphic slope models are of this nature. Certain quantitative assumptions concerning processes operating on slopes are introduced into the model, which then *describes* how the slope develops through time. Or it may have inbuilt constraints and biases with an element of chance controlling its functioning—often by the use of random numbers. Diffusion models frequently operate in this way. They are a useful technique to show, for example, the spread of ideas amongst a population.

Since the end of World War II human geographers have been steadily turning away from regional description, and from the purely factual basis which formed much of the systematic geography before this time. (Physical geographers were never affected in quite the same way, being always more interested in measurement, process, and theory.) Geographers have now moved to a study of process, to a quest for theories which will enable them to explain and make generalizations, and to search for laws which will enable predictions to be made. This change in emphasis, away from a factual study of uniqueness in both the systematic branches and the new areal studies, is fundamental to geography in the second half of this century. It is in this context that the model, as a quantitative technique for formally stating a theory and providing the means to test it, has become important.

V.   PROBABILITY

'Statistics' is a word which has tended to change its meaning in modern times, or at least to be used more loosely. To most people today the meaning of statistics as a branch of mathematics would be taken to include the study of probability theory. This was not the original meaning. A 'statist' was initially one who was skilled in state affairs (i.e. a politician). 'Statistic' was 'that branch of political science dealing with the collection, classification and discussion of facts bearing on the condition of a state or community'; while 'statistics' were 'numerical facts collected and classified' (*Oxford English Dictionary*). Thus 'vital statistics' were merely measurements, from which no inference could be drawn.

The development of probability theory seems to have originated in the late seventeenth century through interest in games of chance. How was a 'fair' wager to be assessed unless some approximation of probability could be calculated? In the simplest example of a game of dice, it is easy to see that a six-faced die has the probability of $\frac{1}{6}$ of landing on any one face, provided there is no bias in the die. That is, at each throw there is one chance in six of any one number lying uppermost. A little thought will show that if two dice are thrown together the probability of both showing the same number or any stated combination of two numbers is $\frac{1}{6} \times \frac{1}{6}$, or 1 chance in 36. Through calculations such as these the study of the mathematical nature of probability originated.

While mathematicians were beginning to think about 'the laws of chance' the statistician was continually searching for ways with which to describe his data, and, as the population grew and technology advanced statistical data increased in volume and became more complex. Sampling became essential, and with it the necessity to determine margins of error, and the extent to which a given sample could be taken as representative of the whole population. Inevitably the old statistics and the new probability theory were interlinked, and became that branch of mathematics which we recognize today simply as 'statistics'.

VI. LEVELS OF MEASUREMENT

Before beginning a study of statistical methods it is first necessary to consider how measurements are made, and it is not always clearly understood that there are four commonly used scales. Each has its own properties, which depend upon the amount of numerical information available. These scales are sometimes referred to as 'levels of measurement'; the more information available the higher the level of measurement that can be achieved.

The lowest level of measurement is when only sufficient information is available to *classify* data, and is usually called the *nominal* scale. Data are sometimes only available in this form. For example, we may study the distribution of villages in a given rural area, decide to classify them either as upland or lowland. We are then able to assign every individual village to one of these two categories. By so doing we have 'measured' the villages on a nominal scale. (In Cyrpus a nominal scale of measurement could result in villages being classified simply as 'Greek' or 'Turkish'.) The villages in each category may then be counted and the total represented by a number. This is termed *counted* data, and is in the form of frequencies (e.g. 10 Greek villages and 13 Turkish).

The next level of measurement is when we have sufficient information not only to establish differences between phenomena (i.e. to place them in different categories) but also to put our data in *rank order*. That is, we know whether one measurement is greater than another, although we do not necessarily know by how much. For example, in conducting a survey in a large urban area it might be desirable to establish in different locations residents' preference for using a particular kind of shop. A sample of people from each location might be asked to state their order of preference for daily shopping in, say, a large departmental store, a self-service supermarket, a smaller shop with service over the counter, or a market. The information obtained will be in rank order only, because it is not reasonable to allocate a precise numerical value to a 'preference'. This is measurement on the *ordinal* scale.

If all the information required for ordinal measurement is present, and in addition the exact interval between each rank is known, then measurement has been achieved on the *interval* scale, and a precise numerical value can be given to each observation. This is the scale we commonly use in recording temperature in degrees centigrade, and because we can measure temperature with precision, we are able to state not only that the mean temperature of one month is higher or lower than that of another, but the exact amount of the difference.

Notice the different *kind* of measurement used with the ordinal and interval scales, and contrast it with the 'counted' data of the nominal scale. On the ordinal and interval scale individual observations are not assigned to categories, but are given *numbers*. Information is thus initially presented as a series of measurements, not frequencies, and may be termed *measured* data (although measurements may, of course, subsequently be placed in groups to form 'group frequencies'). It is necessary that the not always clearly recognized distinction between 'counted' and 'measured' data should be kept in mind.

Finally, if all the requirements for interval measurement are met, but in addition the scale has a true zero, it is termed a *ratio* scale. The property of the scale is that any two measurements bear the same ratio to each other irrespective of the unit of measurement.

The groceries we buy are weighed on a scale with a true zero (where there is no weight), and if we buy two pounds of butter we shall receive twice as much butter as if we bought one pound. Similarly, if we buy two kilos of butter we shall receive twice as much butter as if we bought one kilo. The units of measurement are different, but the *ratio* remains the same. Contrast this situation with temperature measured in degrees centigrade, which is *not* a ratio scale because 0 degrees centigrade is arbitrarily fixed at the freezing-point of water at normal pressure and does not mean that there is no temperature. It is quite incorrect to say that a day with a maximum temperature of 20 degrees centigrade is twice as hot as a day with a maximum temperature of 10 degrees centigrade.

The most rigorous statistical tests employ the interval scale, but, as described later, in order to use them certain stringent requirements must be met, or at least assumed. Very frequently with geographical problems this is not possible, and there is no alternative but to use tests suitable to a lower level of measurement.

The ordinal scale is important to geographers for three reasons: First, data are often only available in ordinal form, i.e. the *amount* of the difference between the ranks may be unobtainable, even when it is meaningful. Secondly, the only possible criterion may be too crude for any precise interpretation to be placed upon it. Thirdly, statistical methods using the ordinal scale, once data have been ranked, are generally much simpler and quicker to calculate than those using interval measurement.

Nominal measurement remains important, because much of the geographer's data is in this form. An example (see p.174) is the 'ordering' of streams in a drainage basin. This division can only be classificatory. A first-order stream is so by definition. It cannot be 'more' or 'less' first order than another. Fortunately for the geographer statistical techniques are available to assist in the analysis of data at this level of measurement.

For many inferential tests employing the ordinal scale it is necessary to assume an underlying continuous distribution. That is, providing it were possible to measure sufficiently accurately, the measurement of any one variate would never be precisely the same as any other. In theory, no two meteorological stations, for example, would record exactly the same annual rainfall, nor the measured angle of two slopes be exactly similar. It is thus theoretically possible to place all sample values in rank order without the occurrence of 'ties'. In practice this frequently does not happen, because our instrumentation is not sufficiently accurate—or, for reasons given above, the allocation of an exact value is not meaningful. We sometimes find therefore that we have more than one number showing the same rank. Fortunately, there are simple ways round this difficulty, which are described at the appropriate places in the text, and the underlying assumption of continuous distribution is preserved.

vii.   GEOGRAPHY AND COMPUTERS

Many geography students now have access to digital computers, and reference to them is sometimes made in the text. There is insufficient space to attempt any instruction in the use of computers, which is outside the scope of this book; instead we simply indicate ways in which they may be usefully employed in solving geographical problems. Output may be in the form of words, sentences, or typed figures produced at typing speed by a teletype machine, or very quickly indeed by a fast line printer. It is possible to programme the computer in such a way that the characters may be located to form a map such as that on page 324, a simple bargraph, or overprinted to provide shading for a simple density map. The computer may also be associated with a graph plotter to provide high quality coloured maps, graphs, and two- and three-dimensional diagrams.

Computers can be used at two quite different levels. There is the person who is thoroughly proficient in one or more of the languages used, who is able to construct programmes to perform a specific task or to read and understand a programme that has been designed by someone else. This is obviously a highly desirable accomplishment. In reality there will always be many students of geography who will not possess the special aptitude this skill requires or, because they are acquiring knowledge in other fields, may not be able to afford the very considerable time and effort necessary to master it. For this reason many departments and computing centres maintain a library of the most commonly required programmes for use with data provided by individual students. Sometimes the computer is operated by specialist staff. Other terminals may be available for student operation, because procedures for feeding in data and programmes are quickly and easily learned.

It cannot be overemphasized that while it may not always be essential for the student to understand the details of how a particular programme operates, it is absolutely essential that he understands the principles of the mathematical or other techniques involved. The role of the computer at this introductory level of quantitative methods is solely to minimize the labour involved in handling large quantities of data, whether for a statistical calculation or some form of model. Problems containing small quantities of data are included to enable the student to understand precisely how a statistic is calculated. For example, the calculation of Kendall's rank correlation coefficient τ (p. 228) is simple enough when only a few pairs of values are involved, but very laborious and with a high probability of error as pairs increase in number. Once the principle is understood there is no point in wasting time working by hand through large samples when the job can be done quickly and accurately by a computer.

Similarly the construction of many models (e.g. the slope model on p. 311) demands far too much calculation time to be worked manually. The use of many models becomes practicable only through the speed of a computer. The programmes for some models, especially simulation models, are relatively complex and often form part of the library stock. It is not necessary for the student to understand the programme, but it is essential that he fully understands the principles upon which the model is constructed. The section on linear programming presents another kind of example, showing how a relatively difficult mathematical technique can be demonstrated using simple graphical methods so that the student can quickly grasp the principle involved. The computer can then provide answers using a large number of variables which would not only take a long time to calculate manually but also require a command of algebra that few students possess.

Finally, the working speed of the computer makes practicable the construction of models which need a large number of 'runs', generally to simulate development through time and space, such as the slope model (p. 311) and the spit simulation model (p. 318). An example of the use of the speed of a computer to make repetitive calculations was given in 1973 by Pinder and Witherick (p. 278), when they used a thousand runs of a programme to establish a formula to determine the mean distance between points randomly distributed along a line. It is a very interesting and simple example of the empirical derivation of a new formula, and a demonstration of the kind of experimental work within the capacity of anyone who has completed this book.

1

SUMMARIZING DATA

1.1   INTRODUCTION

M U C H of the data used by the geographer is comprised of large numbers of numerical values. Take, for example, a series of climatic
statistics for two places given in terms of temperature and rainfall.
Suppose these include the daily maximum and minimum temperatures, and the daily rainfall record over a period of thirty-five years.
We are faced with many sheets of figures for each place, giving such a
profusion of detailed information that direct comparison is impossible. The information has first to be simplified (or summarized) into
a few numbers that measure in some way the various aspects of the
large masses of data in which we are interested. In the case of
climatic statistics rainfall is generally shown as the mean (or average)
annual rainfall over a recent period of years, with possibly the mean
rainfall for each month over the same period. Similarly temperatures
are conventionally summarized to show the mean monthly maximum
and minimum. The mass of detailed climatic data has been reduced
to a few manageable 'representative' figures from which meaningful
comparisons between the two places can be made.

Similarly 'representative' figures may be produced by various
methods to summarize for the purpose of comparison economic and
other data, such as the relative concentration in certain industries of
the working population in different areas. It is with the calculation
of some of these 'representative' figures that this chapter is concerned.

1.2   FREQUENCY DISTRIBUTIONS

Let us first look at the nature of the statistical data which will be
used. It normally consists of a series of figures (or observations) for
one or more categories of phenomena concerned in the subject being
studied. Data are frequently encountered in two different forms.
They might consist of a series of measurements of a particular
phenomenon on a continuous scale (see below): for example, crop
yield per unit area of wheat in different locations. Or it might be
observations of the frequency of an occurrence, either of an object
or a particular value.

Suppose that a study of the industrial economy of an area
included the collection of data on the size of firms in terms of numbers of employees. In all 76 firms were involved and found to vary in

size from 5 to 145 employees. For convenience the data were 'grouped', that is, the value (size) of each firm was placed in a category with a class interval of 10 (i.e. 1–10, 11–20 employees etc.).

It is now necessary to distinguish between variable, variate, and

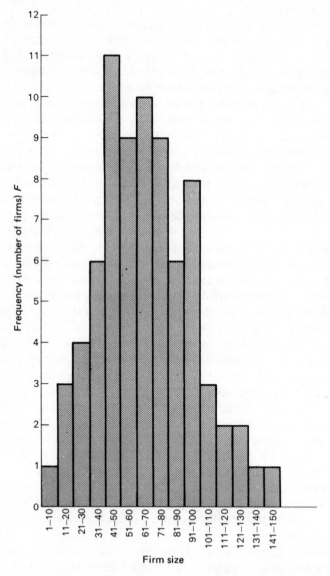

Figure 1.1. Distribution of firms by size grouped into class intervals of 10 employees.

frequency. In this case the *variable* (i.e. a measurable characteristic) is size of firm. The size of each firm is a *variate*, or individual measurement, and normally termed x. The *frequency* of observations having the same value (in this case the number of units falling within the same group) is normally termed f.

The frequency distribution observed in this case was as follows:

Class	1–10	11–20	21–30	31–40	41–50	51–60	61–70	71–80	81–90	91–100	101–110	111–120	121–130	131–140	141–150
f	1	3	4	6	11	9	10	9	6	8	3	2	2	1	1

Figure 1.1 shows this distribution in the form of a histogram, with the class interval plotted on the horizontal axis, and the observed frequency (i.e. the number of firms of sizes falling within each class interval) plotted on the vertical axis. The height of each bar is proportional to the frequency it represents. In this case the frequency distribution deals in whole numbers only. The kinds of data which are presented in indivisible whole units are known as *discrete*. People constitute discrete data: 3·75 employees is an abstract concept which (one hopes) would never form an *observed* frequency, although fractions of people often occur when proportions are considered.

Another type of frequency distribution is that derived from a *continuous* variable. Rainfall recorded at a weather station is an example. It does not rain every day, but it is continuous in the sense that precipitation is measured against a continuous scale on which precision is limited only by the accuracy of the recording instruments. Here a measured day's precipitation of 3·75 cm is not an abstraction, but as exact a statement of quantity as the instruments allow. Temperature has the same property as rainfall, in that it is measured on a continuous scale, but it also differs, because by nature it must form a continuum. (It is not possible for a weather station to have a day without temperature!) A continuous variable therefore can be defined as one which may take any value over a given range, and which may, or may not, be continuous in space and/or time.

1.3 AVERAGES AND MEASURES OF DISPERSION

Probably the most commonly used index to summarize a large number of values is the 'average'. In the newspapers we read about 'average' income, or the 'average' man, and in some textbooks the 'average' yield per acre, or the 'average' size of holding. The 'average' is an extremely useful index. (We all use it constantly in one form or another.) But it is necessary first to decide exactly what we mean. In mathematical language 'average' is synonymous with the 'arithmetic

Data Summary Techniques, showing types of graph, measures of central tendency and dispersion for different levels of measurement found in chapters 1 and 2.

	Appropriate Graphs	Measures of Central Tendency	Measure of Dispersion
Interval and Ratio Scales	DISPERSION DIAGRAM p.6	1. MEDIAN p.5 2. MEAN p.10	1. INTERQUARTILE RANGE p.7 2. STANDARD DEVIATION p.19
Nominal Scales (and grouped interval/ratio data)	1. HISTOGRAM p.2 2. CUMULATIVE FREQUENCY CURVE p.9 3. LORENZ CURVE p.28, 73	MODE p.5	1. LORENZ CURVE INDEX p.28, 75 2. GIBBS MARTIN INDEX p.29 3. GINI COEFFICIENT p.76
Ordinal Scales	Summary techniques are not appropriate for this level of measurement as ranks are evenly distributed between the highest and the lowest values.		

mean', and in this context the latter term is invariably used in this book. Where 'average' is used the less specific meaning is intended, described by the *Shorter Oxford English Dictionary* as 'the medium amount', 'the ruling quantity', or 'the common run'. Under this looser generalization are considered two very useful indexes of 'average', the mode and the median.

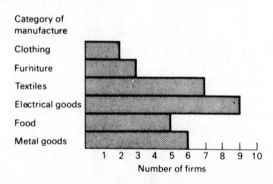

Figure 1.2. Distribution of manufacturing firms classified by types of goods produced

1.3a. | *The mode* |
This may be used when a frequency distribution represents data measured only on a nominal scale (although it may also be applied to other levels of measurement). The modal class is that interval which has the greatest frequency. Figure 1.2 shows a hypothetical distribution of manufacturing firms classified according to the goods they produce. The modal class consists of those firms making electrical goods. (In our previous example (Figure 1.1) the mode is 41 to 50 because there are more firms of this size than any other.) If two categories are very much larger than all others, the distribution may be said to be bi-modal.

1.3b. | *The median* |
The median is a term used for the central value in a series of ordered values. If there is an odd number of ranks the value of the median is that of the central rank. If there is an even number of ranks the median value is the mid-point between the two central ranks. For example, the annual rainfall figures given in Table 1.1 have been drawn on Figure 1.3 in the form if a dispersion graph. It will be seen that the median values for Sydney and Stanley Moor work out at 106·5 cm and 122·5 cm respectively, and that in each case half the annual rainfall totals were greater and half were less than these values.

1.3c. Percentiles
Finding the centre point of a distribution by locating the median value gives us some information about our data, and may be useful for purposes of comparison. What the median does *not* tell us is the *spread* above and below the centre point. Consider the rainfall figures given in Table 1.2. For the period shown Sydney, with a median value of 106·5 cm, had in one year no less than 219 cm of recorded precipitation, and in another only 68 cm. On the other hand for the same period at Stanley Moor, with a median value of 122·5 cm, the highest annual total was 155 cm, and the lowest 103 cm. A glance at Figure 1.3 will show that the *range* is much greater at Sydney, while at Stanley Moor there is less annual variation and the values tend to cluster round the median.

One method of comparing frequency distributions is by the simple visual method of drawing a dispersion graph of the type described above. But visual impressions tend to be vague and indefinite, they vary between individuals, and can be used to study very few distributions at the same time. Some numerical index of dispersion is required to provide an objective method of comparison. One of the simplest is the use of percentiles. It has two advantages. It is quick

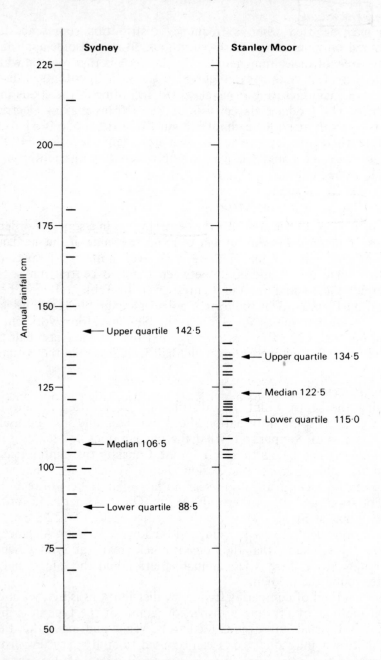

Figure 1.3. The annual rainfall totals for Sydney, Australia, and Stanley Moor, Derbyshire, between 1933 and 1952, drawn in the form of a dispersion graph. Each year is represented by one dash

and easy to calculate, and its meaning is obvious. The most usual index is that of the 25th and 75th percentiles, commonly called the upper and lower quartiles. Both quartiles are indicated on Figure 1.3. The upper quartile is found by taking the 25 per cent highest values (in this case 5) and finding the mid-point between the lowest of these and the next lowest value. Similarly the lower quartile is found by taking the 25 per cent lowest values and finding the mid-point between the highest of these and the next highest value. Between the upper and lower quartiles lies 50 per cent of the values in the distribution, and the difference between the two is termed the *inter-quartile range*. The inter-quartile range is thus a crude index of dispersion. It is at once apparent that the annual precipitation at Stanley Moor, with an inter-quartile range of 19·5, appears on the whole to have much less variation in annual rainfall than Sydney with an inter-quartile range of 54·0. It is necessary to say 'on the whole' if we have the quartile values only, because the extremes at either end of the scale are excluded.

The quartiles are generally found to be the most useful index (and are conventionally most frequently used in comparisons of precipitation), but of course any other percentiles may be obtained to exclude whatever proportion of the rarer, more extreme occurrences is required, e.g. octiles divide a distribution into eight equal groups.

add up 70's.

1.3d. *Cumulative frequency or percentage curves*
Another method, both of depicting a numerical distribution and of deriving percentiles, is the ogive or cumulative frequency curve. This may be illustrated with some age-structure data, which are more usually represented by an age-pyramid (a kind of double histogram— see Figure 1.4), but for which ogives have the following advantages:

   (i) Visual comparison between several age-structures is easier.

   (ii) The shape is not distorted by unequal age-bands (notice the apparently small numbers in the five-year age-bands 0–4 and 5–9 compared with the ten-year band 10–19 in Figure 1.4).

   (iii) Statistics summarizing the age-structure can be quickly read off the ogive.

Figure 1.5 depicts the data in Columns 5 and 6. The gradient of the diagonal (OP) represents the average percentage in each age-band; those parts of a curve which are steeper represent age-bands with above-average percentages, and those which are less steep than the diagonal represent age-bands with below-average percentages. The steeper any segment of a curve, the higher the percentage in the corresponding age-band. In the two graphs drawn, above-average percentages are found in young age-bands, and below-average per-

TABLE 1.1
Age-structure of three national populations

Age group	Grouped percentage				Cumulative percentage below age in last column			
	Brazil 1960		E & W 1861	E & W 1971	Brazil 1960	E & W 1861	E & W 1971	Age
	1	2	3	4	5	6	7	
	M	F	M&F	M&F	1960	1861	1971	
0–4	8·14	7·82	13·46	8·01	15·96	13·46	8·01	5
5–9	7·36	7·13	11·66	8·29	30·45	25·12	16·30	10
10–19	11·12	11·38	20·18	14·24	52·85	45·30	30·54	20
20–29	7·82	8·39	16·71	14·21	69·06	62·01	44·75	30
30–39	6·00	6·12	13·15	11·60	81·18	75·16	56·35	40
40–49	4·36	4·15	10·31	12·45	89·69	85·47	68·80	50
50–59	2·79	2·62	7·29	12·05	95·10	92·76	80·85	60
60–69	1·60	1·54	4·49	10·75	98·24	97·25	91·60	70
70–79	0·58	0·62	2·24	6·18	99·44	99·49	97·78	80
80 & over	0·18	0·24	0·51	2·22	100·	100	100	100

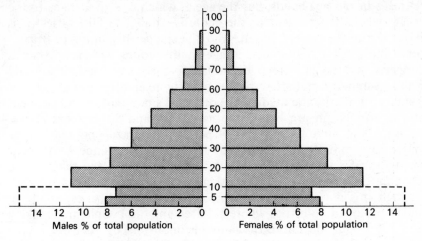

Figure 1.4. Population pyramid of Brazil 1960

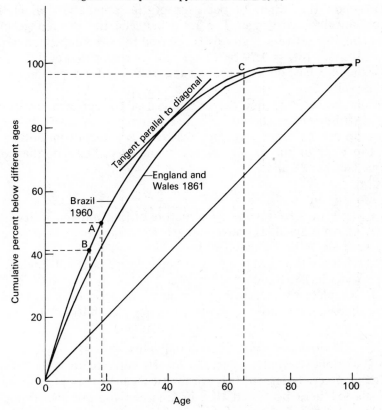

Figure 1.5. Ogives to show age-structures of two national populations

centages in old age-bands, but the age at which one gives place to the other (where the tangent to the curve is parallel to OP) varies—it is about 40 in the case of Brazil. And the steeper beginning to Brazil's curve indicates a higher percentage in the young age group than is indicated by the less steep beginning to the curve for England and Wales. Summary statistics can be quickly read off the curves. For instance, A marks the median age of Brazil's population—50 per cent of the population are below the age of 19 and 50 per cent above. Or again, the curve tells us that 41 per cent of Brazil's population are under 15 (point B) and 3 per cent are 65 and over, since 97 per cent are under 65 (point C). These two age groups are often taken to be 'economically inactive', being either too young or too old to work (though this is only a crude approximation to reality, especially in underdeveloped countries where children and old people may make a significant contribution to the family economy). Together they make up 44 per cent of Brazil's population, and can be regarded as dependent upon the other 56 per cent of the population which is of economically active age. The ratio between the percentage in the inactive and active age groups is referred to as the dependency ratio. For Brazil it is $\frac{44}{56} = 0.79$ or 79 per cent. It means that each person of working age has, on average, to support 0.79 of a dependant as well as himself.

Cumulative frequency curves can be used to represent any numerical distribution. It should be noted that they can have any shape—they do not necessarily get progressively less steep from bottom left to top right of the graph—except in the special case of the Lorenz curve (see pp. 28 and 73).

Consolidation
1. *a.* Draw an ogive for the age-structure of England and Wales, 1971.
   b. At what age does above-average percentage give place to below-average in England and Wales in 1861 and 1971?
   c. What are the median ages and dependency ratios of England and Wales in 1861 and 1971?
   d. Complete the following: 'the more an ogive deviates from the diagonal, the more . . . is the frequency distribution it represents'.

1.3e. *The arithmetic mean* (using individual values)
As stated above what is generally loosely referred to as the 'average' is more precisely termed the 'arithmetic mean', or simply the 'mean'. (It should here be noted that in common use are also found the 'geometric' and 'harmonic' means.) Throughout this book the term 'mean' will refer to the arithmetic mean. If one of the other forms of mean is intended this will be stated. The mean is found by adding

together all the values under consideration and dividing the total by the number of values. For a whole population it is generally denoted by μ, and for a sample by $\bar{x}$. The method of finding the mean may be expressed as

$$\bar{x} = \frac{\Sigma x}{n}$$

Formula 1.1

where x is the variable and n the total number of variates. Σ is simply an instruction to sum all the values of x.

TABLE 1.2
Annual rainfall totals in cm for Sydney (Australia) and Stanley Moor (Derbyshire, England) between 1933 and 1952

	1. Sydney	2. Stanley Moor
1933	108	106
34	165	138
35	79	125
36	77	103
37	132	128
38	99	132
39	85	118
40	100	117
41	68	120
42	123	114
43	129	130
44	79	104
45	180	144
46	92	108
47	105	152
48	99	119
49	168	135
50	219	155
51	135	134
52	150	116
	2392	2498

To find the mean annual precipitation for Sydney and Stanley Moor for the twenty years shown in Table 1.2 we add up the rainfall totals for each year (x) and divide by the number of years (n) in respect of each place.

1. Mean annual rainfall for Sydney

$$\bar{x} = \frac{\Sigma x}{n} = \frac{2392}{20} = 119\cdot6 \text{ cm}$$

2. Mean annual rainfall for Stanley Moor

$$\bar{x} = \frac{\Sigma x}{n} = \frac{2498}{20} = 124 \cdot 9 \text{ cm}$$

Notice that measurement is on an interval scale. Annual rainfall is given to the nearest cm for convenience only. That is, we not only know that more rain fell in one year than another, we also know by precisely how much. Unless interval measurement is achieved we cannot calculate the mean, but have to use some other measure.

1.3f. The arithmetic mean (using grouped data)

Frequently the geographer has occasion to use data which have been grouped into certain arbitrary class boundaries. This may be for convenience when large numbers are involved but also because information is sometimes only available in this form, as frequently is the case with the size of industrial enterprises. Let us suppose we wish, for purposes of comparison, to find the average size of a sample of firms from a specific region by calculating the mean number employed. The information we have is shown in Table 1.3, columns 1 and 3.

TABLE 1.3
Firm size by number of employees

1 Class	2 Class mid-point x	3 Number of firms f	4 Difference from assumed mean d	5 df
1–10	5·5	1	−6	−6
11–20	15·5	3	−5	−15
21–30	25·5	4	−4	−16
31–40	35·5	6	−3	−18
41–50	45·5	11	−2	−22
51–60	55·5	9	−1	−9
61–70	65·5	10	0	0 −86
71–80	75·5	9	1	9
81–90	85·5	6	2	12
91–100	95·5	8	3	24
101–110	105·5	3	4	12
111–120	115·5	2	5	10
121–130	125·5	2	6	12
131–140	135·5	1	7	7
141–150	145·5	1	8	8
				+94
	$\Sigma f = 76$			$\Sigma df = 8$

First we enter the class mid-point for each group. The class interval, for example 11-20, has a class mid-point of 15·5. As the data are grouped we have to assume that the class mid-point is representative of the group. For example, we know there are 11 firms employing between 41 and 50, but we do *not* know how the sizes of the firms are distributed within these limits. By grouping the data and taking the class mid-point as representative we have assumed that sizes fall equally throughout each class. This may not always be so, but generally an excess of high values in some classes will be balanced by excess low values in others, and the element of inaccuracy (which has to be recognized and accepted) will not be great.

The mean of a grouped frequency distribution is normally found by first estimating visually the class near which the actual mean is likely to lie. The class mid-point of that group is then considered to be the 'assumed mean', usually written $\bar{x}_0$. (The use of an assumed mean is not essential to the answer. It is merely a method of reducing the work involved.) It will make no difference to the result how far the assumed mean is from the actual mean, but if the approximation is very inaccurate large numbers will result in the calculation. In the present example the group chosen is 61-70 with a class mid-point of 65·5, which we take as our assumed mean. The frequency f, in this case the number of firms, is then entered for each class. Column 4 is completed to show the difference, in terms of the number of classes, from the assumed mean. Our assumed mean is 65·5 so this class is shown as 0; 55·5 is one class *below* $\bar{x}_0$ and is allocated −1, while 75·5 is one class *above* $\bar{x}_0$ and is allotted 1. Column 5 is the frequency in each class (f) multiplied by the difference (d) of that class from the assumed mean. It now only remains to sum Column 3 (f) and Column 5 (df), and the table is complete.

The formula for finding $\bar{x}$ by using $\bar{x}_0$ is:

$$\bar{x} = \bar{x}_0 + C \cdot \frac{\Sigma df}{\Sigma f} \qquad\qquad \textit{Formula 1.2}$$

where C equals the class interval.

Substituting the figures obtained from Table 1.3

$$\bar{x} = 65·5 + 10 \cdot \frac{8}{76}$$

$$= \underline{66·55}$$

1.3g. Preferred orientation

An average of a different kind is required when we wish to summarize a series of angular measurements. It is a problem which arises, for example, when it is desired to calculate an average value for the orientation of the long axes of a sample of pebbles. Where drift deposits containing pebbles have survived, the direction of the long axes of elongated stones may provide a useful indication of the direction of past ice movements—providing there is a significant average orientation, usually called a preferred orientation, along a given alignment. (We are here concerned to establish a preferred orientation. The question of its significance is dealt with in Chapter 6).

A hypothetical distribution of pebble alignments is shown in Table 1.4a, grouped in classes of 20 degrees. Column 1 shows the class boundaries and Column 3 the number of stones (or frequency 'f') with alignments falling within each class. This information can be displayed visually to indicate frequencies (Figure 1.7) constructed by marking points at a distance from a centre proportional to the number of frequencies in each category and at an angle represented by the mid-point in each class. It will be observed that although orien-

TABLE 1.4a
Stone orientations

1 Deg Class	2 2θ	3 Frequency f	4 Cos 2θ	5 f × Cos 2θ	6 Sin 2θ	7 f × Sin 2θ
0–20	0	22	+1·000	+22·000	0·000	0·000
20–40	40	20	+0·766	+15·320	+0·643	+12·860
40–60	80	12	+0·174	+2·088	+0·985	+11·820
60–80	120	10	−0·500	−5·000	+0·866	+8·660
80–100	160	9	−0·940	−8·460	+0·342	+3·078
100–120	200	7	−0·940	−6·580	−0·342	−2·394
120–140	240	8	−0·500	−4·000	−0·866	−6·928
140–160	280	2	+0·174	+0·348	−0·985	−1·970
160–180	320	15	+0·766	+11·490	−0·643	−9·645
		n = 105		+51·246		+36·418
				−24·040		−20·937
			$\Sigma f \times Cos\,2\theta$	+27·206	$\Sigma f \times Sin\,2\theta$	+15·481

Note. The first value in each class in Column 1 is used to obtain values for Columns 4 and 6, and 10° added to the final answer to give the preferred orientation.

$$\tan 2\bar{\theta} = \frac{\Sigma f \times \sin 2\theta}{\Sigma f \times \cos 2\theta} = \frac{15·481}{27·206} = 0·5690$$

∴ $2\bar{\theta} \simeq 30°$

∴ $\bar{\theta} = 15°$

and the preferred orientation = $15° + 10° = 25°$.

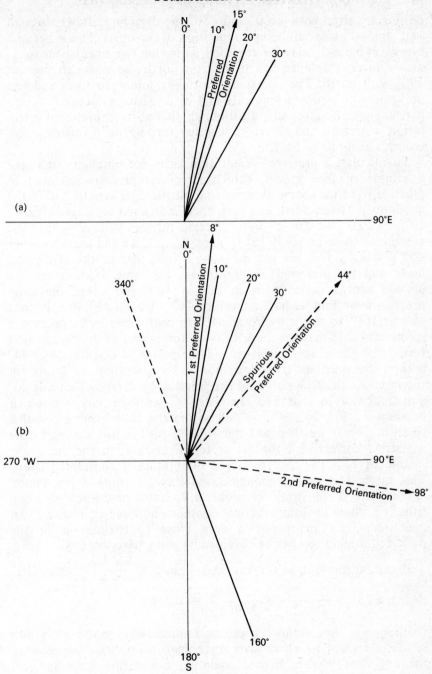

Figure 1.6

tations are given between 0° and 180° the diagram is drawn through 180° to 360°, one half being a repetition of the other. This is because every pebble axis may be regarded as having two angular measurements. For example, an axis at 10° from north also makes an angle of 190° with north. The points are normally joined to form a closed polygon, which provides quickly and easily visual evidence for comparison, and also gives an indication whether a strong preferred orientation is present, and therefore whether time spent in further calculation is likely to be fruitful.

To calculate a preferred orientation let us for simplicity first take a sample of four stones. (Much larger samples are essential in practice.) Let us assume that we find all the long axes lie within the quadrant between north and east, their individual bearings being 0°, 10°, 20°, 30°. We can see by inspection through symmetry that the average of these angles is 15° (Figure 1.6a). If we add them together and divide by four we get the same result, that is, the arithmetic mean and the preferred orientation are the same. If, however, we include another stone with an axis at 160° and adopt the same procedure we find we have a mean of 44° (Figure 1.6b). But the long axis at 160° to north is also at 340° to north, and by inspection it seems probable that the angle of 340° forms part of the distribution formed by the other four axes, and the preferred orientation of 44° is therefore spurious. If, acting upon this assumption, we ignore for the moment the true points of the compass by rotating the scale 20° anti-clockwise so that 340° becomes 0°, we then have a series of bearings 0°, 20°, 30°, 40°, 50°, all within the same quadrant, with a mean of 28°. If we then add this to 340° on the true compass scale we arrive correctly at the 1st preferred orientation for the whole sample of 8°. (The 2nd preferred orientation is considered later.) This method is simple enough when there are only a few values, chosen here to illustrate the problem. It is not practicable in a real situation where the samples could contain a hundred or more angular measurements. Fortunately a very close approximation of the preferred orientation can be obtained by using trigonometry.

It can be shown that for any angle $\tan \theta = \dfrac{\text{Sin } \theta}{\text{Cos } \theta}$. For example,

$$\tan 25° = \frac{\text{Sin } 25°}{\text{Cos } 25°} = \frac{0·4226}{0·9063} = 0·4663.$$

(Although in theory this is a precise figure, owing to the limitations of the tables with which we work it is sometimes only a close approximation.) But we are here considering orientation only and not direction. If we could distinguish between stones pointing south and

and those pointing north it would be appropriate to obtain a mean direction by using vectors, thus:

sum of northerly components of stone axis directions $= \Sigma f \cos \theta$
sum of easterly components of stone axis directions $= \Sigma f \sin \theta$

and therefore $\tan \bar{\theta} = \dfrac{\Sigma f \sin \theta}{\Sigma f \cos \theta}$ where $\bar{\theta}$ is the mean direction of stone axes.

As the long axis of each stone has two diametrically opposed directions (θ and $\theta + 180°$) the sum of northerly and easterly components will cancel each other out and give a zero answer in all cases. (See Appendix 12a.) To use only values of θ between 0° and 180° would mean that there would be a tendency for northerly components to cancel each other out, while easterly components would not. The average orientation would therefore have an easterly bias, because it is expressing an average *direction* rather than an *orientation*. This difficulty can be overcome by using angles of θ between 0° and 180° only, and then doubling them to find an avarage, thus:

$$\tan 2\bar{\theta} = \frac{\Sigma f \sin 2\theta}{\Sigma f \cos 2\theta}$$

Applying this technique to the data in Figure 1.6b (Table 1.4b).

TABLE 1.4b

θ	2θ	Cos 2θ	Sin 2θ
0	0	+1·0000	0·0000
10	20	+0·9397	+0·3420
20	40	+0·7660	+0·6428
30	60	+0·5000	+0·8660
160	320	+0·7660	−0·6428
		+3·9717	+1·2080

and $\tan 2\bar{\theta} = \dfrac{\Sigma \sin 2\theta}{\Sigma \cos 2\theta} = \dfrac{1 \cdot 2080}{3 \cdot 9717} = 0 \cdot 3041$ therefore $2\bar{\theta} = 16°$ and $\bar{\theta} = 8°$, which is what we found arithmetically.

It should be noted that for any value of $\tan 2\theta$ there will be two values 180° apart. Thus if $\tan 2\bar{\theta} = +1$, $2\bar{\theta}$ is either 45° or 225°. If $\tan 2\bar{\theta} = -1$, then $2\bar{\theta}$ is either 135° or 315°. Therefore there will always be two values of $\bar{\theta}$. In the example given above (Table 1.4b and Figure 1.6b) $\bar{\theta}$ may be either 8° (1st preferred orientation) or 98° (2nd preferred orientation). This reflects the fact that any group of *orientations* (as opposed to directions) will have two preferred alignments at right angles to each other. The only way to determine which is appropriate for the data in question is by inspection. In Figure 1.6b 8° is clearly more appropriate as a 'preferred angle' than 98°. The more appropriate angle of the two is called simply 'the preferred orientation'.

As with all averages, a preferred orientation may be misleading. A group of stones may have a strongly bi-modal distribution with axis alignments falling into two main groups, half NE/SW and half NW/SE. The preferred orientation would then be either N./S. or E./W.; a meaningless result, unless from other geomorphological evidence it could be shown that the modes were not part of the same population, but represented two distributions, possibly related to ice movement in two different directions—in which case two separate calculations would be required, and each sample would have a single mode.

Table 1.4a shows the completed calculation of preferred orientation for the sample of 105 elongated stones. In this case the orientations of stone axes were grouped into 20° classes and cos θ and sin θ were taken for the smallest angle of each class. 10° was therefore added to $\bar{\theta}$ to bring the answer to the class mid-point. The choice of preferred orientation is here not in doubt, as Figure 1.7 shows a clear N.E./S.W. alignment.

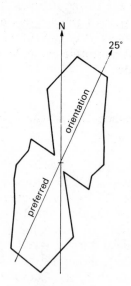

Figure 1.7

1.3h. The standard deviation (individual values)

This is probably the single most important of all statistical 'yard-sticks', because (i) it is a measure of dispersion in which *all* values are taken into account, (ii) it has many uses in addition to (i), (iii) it is necessary for the probability theory outlined in Chapter Four, and the inferential techniques which follow. For these reasons it is absolutely essential that a thorough understanding should be achieved at this stage.

The standard deviation is a measure of the *dispersion* of a series of given values from their arithmetic mean. It is calculated first by obtaining the arithmetic mean, and then by measuring how much each value differs from it. Those values higher than the mean will result in positive differences, while those lower will be negative. Each difference is squared. The squares are summed and the total divided by the number (n) of values (x). This figure is known as the *variance* of the distribution. The standard deviation (normally represented by little sigma (σ) in the case of a whole population, and s in the case of a sample) is obtained by taking the square root of the variance. The formula for the standard deviation which instructs us to carry out these operations is:

$$s = \sqrt{\left[\frac{\Sigma(x - \bar{x})^2}{n}\right]}$$ *Formula 1.3*

$(x - \bar{x})$ tells us to subtract the mean ($\bar{x}$) from each value of the distribution in turn; $(x - \bar{x})^2$ to square each difference. $\Sigma(x - \bar{x})^2 /n$ tells us to divide the sum of the squares by the total number (n) of the values (x) involved, and $\sqrt{[\ \Sigma(x - \bar{x})^2 /n]}$ to take the square root of the final result. A simple example will serve to show how this calculation is best made.

Example

Calculate the standard deviation of the following:

  1 2 3 4 5 6 7 8 9

First we must find the mean, in this case $\bar{x} = 5$.
Next tabulate the data. (see overleaf)

Now substitute the appropriate figures in the formula for the standard deviation

$$s = \sqrt{\left[\frac{\Sigma(x - \bar{x})^2}{n}\right]}$$

$$= \sqrt{\frac{60}{9}} \qquad = \underline{2 \cdot 58} \ldots$$

TABLE 1.5

x	$x - \bar{x}$	$(x - \bar{x})^2$
1	−4	16
2	−3	9
3	−2	4
4	−1	1
Mean = 5	0	0
6	1	1
7	2	4
8	3	9
9	4	16
$\Sigma x = \overline{45}$		$\Sigma(x - \bar{x})^2 = \overline{60}$ $n = 9$

The definition of the standard deviation is that it is *the root mean square deviation from the mean*.

The above example was chosen to demonstrate the nature of the standard deviation in the simplest possible way. It was so arranged that the mean would be an integer (a whole number). But in practice the mean is rarely an integer and the calculation of each difference involves us in decimal places. These awkward numbers then have to be squared and summed, involving time, labour, and the possibility of error. Formula 1.3 was used in the above example because it explains how the standard deviation is arrived at. Fortunately, there is an alternative, Formula 1.4a, which gives exactly the same result (because it is derived algebraically from Formula 1.3)[2] although easier to work out,

$$ s = \sqrt{\left\{ \frac{\Sigma x^2}{n} - \left(\frac{\Sigma x}{n} \right)^2 \right\}} . \qquad \textit{Formula 1.4a} $$

It will be observed that $(\Sigma x/n)$ is the arithmetic mean, so Formula 1.4a is frequently written

$$ s = \sqrt{\left\{ \frac{\Sigma x^2}{n} - \bar{x}^2 \right\}} . \qquad \textit{Formula 1.4b} $$

Example

Calculate the standard deviation of the annual precipitation totals for Sydney and Stanley Moor between 1933 and 1952. (These data have been used again so that direct comparison can be made with the techniques discussed previously in 1.3c.)

TABLE 1.6
Annual precipitation in cm—1933 to 1919

Stanley Moor		Sydney	
x	x^2	x	x^2
106	11 236	108	11 664
138	19 044	165	27 225
125	15 625	79	6 241
103	10 609	77	5 929
128	16 384	132	17 424
132	17 424	99	9 801
118	13 924	85	7 225
117	13 689	100	10 000
120	14 400	68	4 624
114	12 996	123	15 129
130	16 900	129	16 641
104	10 816	79	6 241
144	20 736	180	32 400
108	11 664	92	8 464
152	23 104	105	11 025
119	14 161	99	9 801
135	18 225	168	28 224
155	24 025	219	47 961
134	17 956	135	18 225
116	13 456	150	22 500
2498	316 374	2392	316 744

$$s = \sqrt{\left\{ \frac{\Sigma x^2}{n} - \left(\frac{\Sigma x}{n}\right)^2 \right\}} \qquad\qquad s = \sqrt{\left\{ \frac{\Sigma x^2}{n} - \left(\frac{\Sigma x}{n}\right)^2 \right\}}$$

$$= \sqrt{\left\{ \frac{316\,374}{20} - \left(\frac{2498}{20}\right)^2 \right\}} \qquad = \sqrt{\left\{ \frac{316\,744}{20} - \left(\frac{2392}{20}\right)^2 \right\}}$$

$$= \underline{14{\cdot}8} \ldots \qquad\qquad\qquad = \underline{39{\cdot}2} \ldots$$

It will be noted that the mean precipitation at Stanley Moor (124·9 cm) was only slightly higher than at Sydney (119·6 cm), but that the standard deviation for the former (14·8) is very much less than the latter (39·2). An understanding of the significance of the standard deviation as a measure of dispersion is clearly apparent from Figure 1.8. Notice how the deviations of the annual totals for Stanley Moor cluster closely round the mean, whereas those for Sydney fluctuate very considerably. The standard deviation is a precise arithmetical measure of those fluctuations.

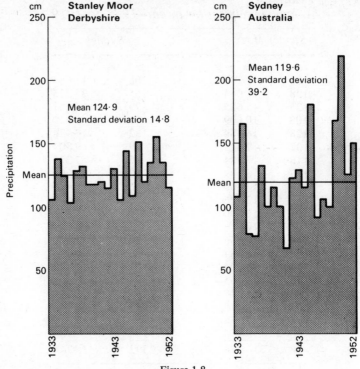

Figure 1.8

Summary. The standard deviation (individual values)
1. Tabulate the values (x) and their squares (x^2) as in Table 1.6.
2. Find the mean $(\bar{x})$ of the values of x and square it.
3. Sum values of x^2 and divide by n (the number of values).
4. Subtract $\bar{x}^2$ from the result of 3.
5. Obtain the square root of the result of 4.

1.3i.   The standard deviation (grouped frequencies)
The calculation of the standard deviation for grouped frequencies differs from the previous calculation of the standard deviation in two respects. The deviations this time are in terms of class intervals from the assumed mean; and the result has to be multiplied by the value of the class interval employed. The example below uses the same firm-size data given previously on page 12.
    The standard deviation is obtained from the formula:

$$s = C. \sqrt{\left\{ \frac{\Sigma d^2 f}{\Sigma f} - \left(\frac{\Sigma df}{\Sigma f}\right)^2 \right\}} \qquad \textit{Formula 1.5}$$

where C is the amount of the class interval.

TABLE 1.7

1	2	3	4	5	6
				Difference from assumed mean	
Class	Class mid-point	Number of firms			
	x	f	d	df	d^2f
1–10	5·5	1	−6	−6	36
11–20	15·5	3	−5	−15	75
21–30	25·5	4	−4	−16	64
31–40	35·5	6	−3	−18	54
41–50	45·5	11	−2	−22	44
51–60	55·5	9	−1	−9	9
61–70	65·5	10	0	0	0
71–80	75·5	9	1	9	9
81–90	85·5	6	2	12	24
91–100	95·5	8	3	24	72
101–110	105·5	3	4	12	48
111–120	115·5	2	5	10	50
121–130	125·5	2	6	12	72
131–140	135·5	1	7	7	49
141–150	145·5	1	8	8	64
		$\Sigma f = 76$		$\Sigma df = 8$	$\Sigma d^2f = 670$

Therefore

$$s = 10 \sqrt{\left\{ \frac{670}{76} - \left(\frac{8}{76}\right)^2 \right\}}$$

$$= \underline{29\cdot 67 \ldots}$$

Summary. The standard deviation (grouped frequencies)

1. Decide upon the class interval.

2. Tabulate the data as in Table 1.7 showing the class intervals, class mid-points, and the frequencies (*f*) with which values occur in each class.

3. By inspection guess within which class you think the mean is likely to lie. The mid-point of that class will then be your assumed mean.

4. Note the difference (*d*) in terms of class interval between each class and the assumed mean (Column 4), and multiply each difference by the frequency for that class (Column 5).

5. Square the difference of each class from the assumed mean (d^2) and multiply by the frequency for that class.

6. Sum *f*, *df*, and d^2f and substitute in Formula 1.5.

1.3j. The coefficient of variation

This is an expression of variation (another way of indicating dispersion) obtained by converting the standard deviation to a percentage of the mean. For example, using the annual precipitation figures for Stanley Moor and Sydney, we can calculate the coefficient of variation (V) for each:

$$V. \text{ Stanley Moor} = \frac{s}{x} \times 100 = \frac{14 \cdot 8}{124 \cdot 9} \times 100 = 12\% \text{ approx.}$$

$$V. \text{ Sydney} = \frac{s}{x} \times 100 = \frac{39 \cdot 2}{119 \cdot 6} \times 100 = 33\% \text{ approx.}$$

This technique is useful when it is desired to plot variability on a map, because variations appear as readily comparable percentage differences. A map based on the coefficient of variation of annual rainfall over the British Isles was prepared by Gregory (1963), and presents an interesting pattern, showing considerable local variation in Britain. (It is worth noting that in most countries the authorities responsible for public water supply operate a large number of their own rain gauges, and are normally prepared to make available detailed statistical information.)

Consolidation

2. The mean annual totals of precipitation at Chatsworth Gardens, Derbyshire, from 1878 to 1966 (years 1882–4, 1886, 1924, 1961 are missing), correct to the nearest centimetre, are as follows:

1878–1902	1903–1923	1925–1945	1946–1966
111	102	86	102
109	71	78	85
127	75	95	85
119	87	85	72
83	93	75	80
52	75	96	99
74	85	70	74
89	92	77	67
75	69	64	102
102	122	62	66
85	91	86	90
72	104	82	80
86	103	85	100
77	96	70	69
77	74	86	108
82	79	84	73
76	75	85	76
82	71	69	73
96	50	71	106
79	69	94	101
75	81	71	

Calculate the inter-quartile range and standard deviation for the years 1946–66.

3. Convert the annual totals 1878–1966 into frequencies using a class interval of 5, and express in the form of a histogram (bargraph).

4. Calculate the standard deviation for the grouped frequencies. Calculate and compare the coefficients of variation for the years 1878–1902 and 1925–45.

1.4 THE LORENZ CURVE

Sometimes the geographer is engaged in the comparison of the industrial structure of one town or region with that of another in terms of the numbers employed in each occupation. One aspect of industrial structure might well be the dependence of such a town on a particular manufacture, or the extent to which employment was distributed among a variety of industries. One method of comparison which is being increasingly used is the Lorenz curve. This is particularly useful because it provides a visual comparison of differences, and additionally a precise index of the difference can be calculated from the information used in drawing the curve.

The Lorenz curve may be used to measure industrial diversification by the extent to which a given distribution differs from a hypothetical even distribution. It is drawn by expressing the frequency in each category (in our case employment in each manufacturing industry) as a percentage of the total frequencies, and then plotting the result graphically in the form of a cumulative percentage curve.

Table 1.8 is an example of the tabulated data. Similar calculations were made for Nottingham, and for England and Wales as a whole. From these cumulative percentage figures the curves on Figure 1.9 were constructed. It is evident that if every category had exactly the same percentage of the total employed (i.e. if employment were absolutely evenly distributed), the cumulative curve would be a straight line. The *visual effect of concentration* is shown as the area contained between the curve and the line of even distribution increases in size. From Figure 1.9 it is immediately apparent that fewer industries account for a higher percentage of employment in Coventry than in Nottingham, and much more so than in England and Wales as a whole.

The Lorenz curve has many applications in geography. (It is, for example, a useful technique for comparing the relative concentration or dispersion of population areally. See Chapter Two.) But, as always, caution must be observed in interpreting the results, especially when using the Standard Industrial Classification Orders of the National Census. A high proportion of workers may be concentrated in industries covered by SIC Order VI Engineering and Electrical Goods, a general term which obscures the fact that no fewer than 20 different kinds of manufacture are included in that category varying from guns and industrial plant to watches and surgical instruments. (The 1971 Census used a new S.I.C. which divided Engineering into three industrial orders.)

TABLE 1.8
Selected data from industry tables, 1961. Census of Population

SIC Order	Rank	Employed	Percentage of total	Cumulative percentage
Coventry	1			
VIII	1	59 550	53·63	53·63
VI	2	33 130	29·83	83·46
X	3	7 570	6·81	90·27
IX	4	4 960	4·47	94·74
XV	5	1 340	1·21	95·95
XIV	6	1 190	1·07	97·02
III	7	1 020	0·92	97·94
V	8	840	0·76	98·70
XIII	9	740	0·67	99·37
XVI	10	320	0·29	99·66
IV	11	250	0·23	99·89
XII	12	110	0·10	99·99
XI	13	10	0·01	100·00
VII	14	0	0·00	100·00
		111 030	100·00	1310·62
			Approx. total	1311
Nottingham				
X	1	21 670	25·83	25·83
III	2	16 770	19·99	45·82
VI	3	11 940	14·24	60·06
XII	4	9 680	11·54	71·60
VIII	5	8 770	10·46	82·06
XV	6	5 410	6·45	88·51
IV	7	3 430	4·09	92·60
XIV	8	2 760	3·29	95·89
IX	9	990	1·18	97·07
XIII	10	830	0·99	98·06
V	11	730	0·87	98·93
XVI	12	470	0·56	99·49
XI	13	310	0·37	99·86
VII	14	120	0·14	100·00
		83 880	100·00	1155·78
			Approx. total	1156

TABLE 1.8—*continued*

SIC Order	Rank	Employed	Percentage of total	Cumulative percentage
England and Wales				
VI	1	1 870 760	24·40	24·40
VIII	2	805 670	10·51	34·91
X	3	692 010	9·02	43·93
III	4	620 900	8·10	52·03
V	5	563 030	7·34	59·37
XV	6	548 340	7·15	66·52
XII	7	517 670	6·75	73·27
IX	8	502 840	6·56	79·83
IV	9	463 410	6·04	85·87
XIII	10	300 940	3·92	89·79
XIV	11	276 510	3·61	93·40
XVI	12	276 070	3·60	97·00
VII	13	173 540	2·26	99·26
XI	14	56 310	0·74	100·00
		7 668 000	100·00	999·58
			Approx. total	1000

Comparison of Lorenz curves with the straight line of even distribution is a quick visual means of describing (in this particular example) industrial diversification. But visual interpretation is inevitably only a rough approximation. A numerical index gives greater precision, and takes up much less space. Such a figure may be derived from the graph itself. It is the ratio of the area of the right-angled triangle with the base line of even distribution, and the area contained by individual curves. This is perfectly satisfactory, although calculation of the area contained by the curves is time-consuming and awkward.

Fortunately it is possible to work directly from the information already tabulated to produce a slightly different and in some ways more useful figure. This is obtained by first summing the cumulative percentage totals. These are informative in themselves and constitute a kind of crude index. Study of Table 1.8 discloses that the greater the concentration of frequencies in a small number of categories the higher the sum of the cumulative percentage will be. Thus in our example, Nottingham with a total of 1156 shows considerably more diversification than Coventry with a total of 1311, compared with a figure of about 1000 for England and Wales as a whole. Absolute specialization, i.e. the concentration of *all* frequencies in one category, with 14 categories would result in a total of 1400 (100 per cent in rank one and 100 per cent for each subsequent rank). The crude

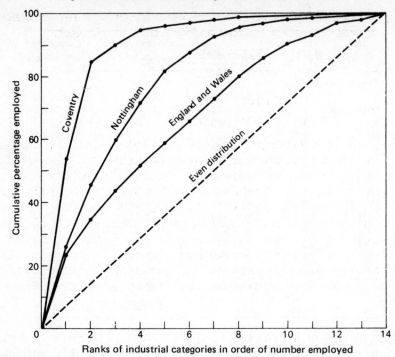

Figure 1.9. Lorenz curves comparing the distribution of employment by occupations between Coventry, Nottingham, and England and Wales.

Categories of employment (like those in Table 1.8) are the Standard Industrial Classification Orders III to XVI used in the 1961 Census.

index therefore is a measure of the extent to which any area, including the whole region, differs from absolute specialization (or concentration of frequencies) into one category.

A refinement of the crude index is to compare smaller areas, such as towns, counties, or sub-regions, with some larger region in which they are situated. In our example the diversification of manufacture in Coventry and Nottingham might more realistically be compared with the *existing* structure in England and Wales rather than to some hypothetical extreme. This may be done by again using the cumulative totals. The index may be found by

$$I = \frac{A - R}{M - R}, \qquad\qquad Formula\ 1.6$$

where I is the index of concentration, A is the area cumulative percentage total, R is the regional cumulative percentage total, and M is the maximum cumulative percentage total assuming 100 per cent of the frequencies in Rank 1.

In our example R is 1000 and M, because we have 14 categories, is 1400. Indexes for Nottingham and Coventry are therefore

$$I \text{ Nottingham} = \frac{1156 - 1000}{1400 - 1000} = 0.39 \ldots$$

$$I \text{ Coventry} = \frac{1311 - 1000}{1400 - 1000} = 0.78 \ldots$$

Similarly economic regions within the country might be compared with the United Kingdom as a whole. Based on the same classification the index for the Yorkshire and Humberside region is 0·14 . . .

An index figure of 1·0 represents an 'absolute' concentration of all frequencies into one category. It should be remembered that an index approaching 0 does *not* mean (in this case) absolute diversification, but a distribution of frequencies in a particular area *similar to that for the whole region*. In this sense the index is a good measure of the extent to which any one area has the same level of concentration as a whole country or region in respect of a particular distribution.

1.5. THE GIBBS-MARTIN INDEX OF DIVERSIFICATION

A useful alternative index for testing the diversification of employment in industry (but capable of use in other fields) was developed by Gibbs and Martin (1962). If the labour force in a region is concentrated wholly in one industry the index is zero; if it is evenly distributed throughout every industry (i.e. maximum diversification) the index approaches 1.

The formula is $1 - \Sigma x^2/(\Sigma x)^2$, where x is the number of employees in each industrial category. *Formula 1.7*

Clements and Sturgis (1971) give a hypothetical example showing complete diversification and concentration in a paper entitled 'Population Size and Industrial Diversification' (see overleaf).

Besides providing a useful means of comparing regional differences, this has an advantage over other indexes in that it is relatively easy to calculate, *because it avoids the need to reduce actual numbers to percentages.* It should be noted, however, that (like the Lorenz curve) the index value is related partly to the number of categories used. Clements and Sturgis note that with four categories the maximum value obtained for absolute diversification is 0·7500, while for ten categories it is 0·9000. More generally, for n categories, the

	Industrial Employment	
Industry	Community A	Community B
1	12 000	1000
2	–	1000
3	–	1000
4	–	1000
5	–	1000
6	–	1000
7	–	1000
8	–	1000
9	–	1000
10	–	1000
11	–	1000
12	–	1000
Σx	12 000	12 000
Σx^2	144 000 000	12 000 000
$(\Sigma x)^2$	144 000 000	144 000 000
$1 - \Sigma x^2/(\Sigma x)^2$	0·0000	0·9167

maximum score is $\dfrac{n-1}{n}$. For the index to be valid the same *number* of categories should be used whenever comparison is made.

An actual example employs the index to test diversification of exports. In this case the value of the exports is used, divided into the ten Standard International Trade categories as published by the United Nations in the *Yearbook of International Trade Statistics, 1965*. It should be noted that the currency, and the value of the currency in terms of U.S. dollars or gold, is immaterial. Provided that the same currency is used for every category in each country (as given by the U.N.), it is perfectly valid to compile the index for the U.K. in pounds sterling and for West Malaysia in Malayan dollars. The reason for this is that the frequencies (in terms of the local currency) of categories are being compared within individual countries, and there is no comparison in terms of value between one country and another.[3]

More than usual caution must be observed in assessing the significance of the index, because, as in 1.4 above, the immense diversity of world trade has been reduced to only ten categories. For example, SITC Code 6 (manufactured goods classified by material) for the U.K. contains 36 sub-headings.

The data shown in Table 1.9 are divided into ten categories, the range of the Gibbs-Martin index will therefore run from 0 (absolute concentration) to 0·9 (completely even distribution). The United

TABLE 1.9
Calculation of the Gibbs-Martin index using export figures, 1965

United Kingdom

SITC Code	Exports (x)	x^2	
0	£ 154·76 million	23 950·66	$I = 1 - \dfrac{\Sigma x^2}{(\Sigma x)^2}$
1	143·16	20 494·79	
2	144·26	20 810·95	
3	133·42	17 800·90	$= 1 - \dfrac{5\ 823\ 825}{21\ 941\ 636}$
4	6·70	44·89	
5	439·14	192 843·94	
6	1210·76	1 465 939·78	$= 0.74 \ldots$
7	1985·62	3 942 686·78	
8	356·68	127 220·62	
9	109·69	12 031·90	
	4684·19	5 823 825·21	

$$(\Sigma x)^2 = 21\ 941\ 636$$

Libya

0	£(Libyan)	0·190 million	0·036	$I = 1 - \dfrac{\Sigma x^2}{(\Sigma x)^2}$
1		0·041	0·002	
2		1·448	2·097	
3		280·326	78 582·670	$= 1 - \dfrac{78\ 584}{79\ 528}$
4–9		0·003	0·000	
		282·008	78 584·805	$= 0.01 \ldots$

$$(\Sigma x)^2 = 79\ 528·512$$

West Malaysia

0	£(Malayan)	149·51 million	22 353·24	$I = 1 - \dfrac{\Sigma x^2}{(\Sigma x)^2}$
1		1·76	3·10	
2		1085·26	177 789·27	
3		50·20	2 520·04	$= 1 - \dfrac{2\ 082\ 506}{6\ 264\ 408}$
4		122·57	15 023·40	
5		44·23	1 956·29	
6		925·64	856 809·41	
7		66·32	4 398·34	
8		26·87	722·00	$= 0.67 \ldots$
9		30·52	931·47	
		2502·88	2 082 506·56	

$$(\Sigma x)^2 = 6\ 264\ 408·29$$

TABLE 1.9–*continued*

SITC Code		*Australia*		
0	$(A)	1023·52 million	1 047 593·19	$I = 1 - \dfrac{\Sigma x^2}{(\Sigma x)^2}$
1		8·53	72·76	
2		989·72	979 545·68	
3		83·43	6 960·56	
4		17·55	308·00	$= 1 - \dfrac{2\,126\,912}{7\,037\,082}$
5		52·10	2 714·41	
6		265·65	70 569·92	
7		120·63	14 551·60	$= \underline{0\cdot70} \ldots$
8		31·68	1 003·62	
9		59·94	3 592·80	
		2652·75	2 126 912·54	

$$(\Sigma x)^2 = 7\,037\,082\cdot56$$

Kingdom ($I = 0\cdot74$) is thus seen as having a relatively diversified export trade, despite the high values for SITC 6 and 7 (manufactured goods classified by material and machinery and transport equipment) which one would expect in a highly industralized state. West Malaysia ($I = 0\cdot67$) has a rather less diversified export trade with high values in SITC 2 (mainly in natural rubber), and SITC 6 (mainly in unwrought tin), typical of a developing country and primary producer. Libya ($I = 0\cdot01$) has been included as an extreme example, similar to other countries with large oil reserves, where dependence on a single export (crude petroleum) is almost complete.

With $I = 0\cdot7$ it would seem that Australia has a relatively diversified export trade, although with high values in categories 0 and 2. In fact, an examination of the export figures in more detail reveals that categories 0 (Food) and 2 (crude materials) may to some extent be dependent upon each other. For example, a high proportion of SITC 0 is meat and dairy products, and a high proportion of SITC 2 is wool and hides. This is not an attempt to interpret the figures, but simply to indicate the necessity for care.

An obstacle to the use of the index is the awkwardness of the very large numbers which result from using the full published figures, particularly if a calculator is not available to do the work. Fortunately, the totals may be simplified without, in most cases, materially altering the answer. The index is reworked below for Australia and the United Kingdom with simplified figures. In the case of Australia the published figures are in millions of Australian dollars correct to two places of decimals. This has been reduced to hundreds of millions correct to one place of decimals; x^2 is correct to the nearest whole number. Figures for the United Kingdom have similarly been adjusted to hundreds of millions of pounds sterling correct to one decimal place. The calculation is now quick and easy.

	Australia 100m (A)\$		U.K. £100m	
SITC	x	x^2	x	x^2
0	10·2	104	1·5	2
1	0·1	0	1·4	2
2	9·9	98	1·4	2
3	0·8	1	1·3	2
4	0·2	0	0·6	0
5	0·5	0	4·4	19
6	2·7	7	12·1	146
7	1·2	1	19·9	396
8	0·3	0	3·6	13
9	0·6	0	1·1	1
	$\Sigma x = 26·5$	$\Sigma x^2 = 211$	$\Sigma x = 47·30$	$\Sigma x^2 = 583$

$$(\Sigma x)^2 = 702 \qquad\qquad (\Sigma x)^2 = 2237$$

$$I = 1 - \frac{\Sigma x^2}{(\Sigma x)^2} \qquad\qquad I = 1 - \frac{\Sigma x^2}{(\Sigma x)^2}$$

$$= 1 - \frac{211}{702} \qquad\qquad = 1 - \frac{583}{2237}$$

$$= \underline{0·70 \ldots} \qquad\qquad = \underline{0·74 \ldots}$$

Because of the influence of the larger numbers I Aust and I UK are exactly the same figures that we got before, in spite of the fact that in the process of squaring some categories were reduced to 0. If the same procedure were adopted in the extreme case of Libya the index would change from 0·01 to 0. Generally speaking, with an admittedly crude index of this kind, fine margins of accuracy are worth sacrificing for speed and ease of calculation.

It will be observed that the index derived from Lorenz curve data is described as an index of concentration, while the Gibbs-Martin index is called an index of diversification, although each obviously is an index of both extremes. The difference lies in the fact that the former gives a high numerical value for concentration, and the latter a high numerical value for dispersion. The Lorenz curve index is therefore treated as an index of *concentration*, and the Gibbs-Martin index one of *diversification*.

1.6. THE TERNARY DIAGRAM

So far we have dealt with situations in which a large number of statistics *of the same kind* have been reduced to single indexes of description. These are useful, but more often than not the problem is concerned with the relationships between more than one component or more than one aspect of the same component. One of the clearest ways of describing the relationship of two components is by means of a scattergram with scales located along the horizontal (abscissa) and vertical (ordinate) axes. Population growth through time is often described in this way. An adaptation of the scattergram, called a ternary diagram (or triangular graph) is a device for plotting data simultaneously on three scales. It may be used whenever scores on the three scales add up to 100—i.e. are percentages, and is an objective method of grouping phenomena in accordance with the relative dominance of each. The diagram consists of an equilateral triangle, the sides of which form three scales, graduated 0 to 100 per cent, so that each apex forms 0 on two scales and 100 per cent on the other. From each scale lines at an angle of 60° are drawn to carry values on that scale across the graph. Provided that the three scores total 100 per cent, they may be represented by a single point on the diagram. The position of any point within the triangle reflects the relative dominance of each component. Soils are sometimes classified by the texture of the inorganic fraction, that is, according to the proportion of clay, silt, and sand which they contain. A ternary diagram is frequently used to portray visually the relative proportion of each constituent in a number of soil samples, to demonstrate how they may be grouped under certain general headings.

Figure 1.10a is an example of such a diagram. Interpretation should be made with care until the reader is practised in the use of the technique. Figures 1.10b and 1.10c are of great help in understanding the significance of the location of any point—a method used by G. C. Dickinson (1970). It will be seen that the ten soil samples fall into four main groups, but though the texture of the soil in each group is generally similar, each sample, with the exception of the centre group, is not exactly the same as the other members of the group. Examine the location of samples 1, 2, and 3. All three fall within the area where the silt fraction is greater than the sand and clay fractions added together. (That is $B > A + C$.) All may therefore be classified as silts. But the position of 1 also shows that the sand fraction is greater than the clay. ($C > A$.) So 1 might be called a silt loam. Sample 3 on the other hand has a higher proportion of clay than sand ($A > C$) and might be termed a silty clay. However, sample 2 in addition to silt has an equal amount of sand and clay, and might be called a silty clay loam. Samples 7 and 8 have roughly equal proportions of all three ingredients and might constitute a clay loam.

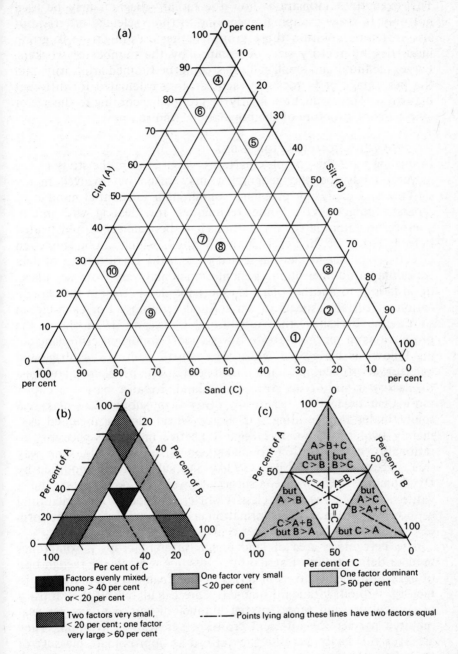

Figure 1.10. A ternary diagram describing ten soil samples classified according to the proportions each contains of sand, salt, and clay

This example demonstrates how the ternary diagram may be used not only to show groupings according to three selected criteria, but also differences within those groups. Other uses might be to group industries by factory size, determined by the number of workers. Large, medium, and small factories could be defined arbitrarily, and the percentage of factories in each category calculated for different industries. Towns might similarly be grouped according to the age or employment structure of the population. And so on.

1.7. WEAVER'S COMBINATION INDEX

In dealing with the complex patterns which arise in the study of the cultural landscape the geographer may often be involved in describing and analysing problems containing a substantial number of separate components. Some techniques for dealing with mutli-component interrelated distributions have become very sophisticated indeed. There are, however, simple methods which can be employed effectively by the student as a first step in the classification of data containing a number of components which may vary from one place, or situation, to another. This is especially clearly seen in the classification of certain aspects of agricultural economy. For example, in south-west Berkshire the Agricultural Census of 1958 shows the proportions in terms of area of the six most important crops (including fallow) as barley 45, wheat 21, oats 12, fallow 8, cabbage and kale for stock feeding 7, and mixed corn 2. The problem is how best to classify a multi-crop area of this kind. Because barley occupies more than twice the area of any other crop, should it be classified solely under that heading? Or should wheat also be included as a mainly two-crop economy? Some objective method is necessary to rationalize categorization to make comparison possible. One way was suggested by J. Weaver (1954), and subsequently modified by D. Thomas (1963). Weaver's suggestion was for a combination index which could be used to classify different aspects of agricultural economics, provided the information could be converted to the form of proportions totalling 100 per cent.

The percentage area of land under different crops is sometimes used to define types of agriculture. In some respects the recognition of agricultural enterprises in terms of the 'man-days' is a more satisfactory method. 'Standard man-days' are calculated by the Ministry of Agriculture and represent the number of working days per year required per acre by different crops, or per head of livestock. They are a useful device because they afford an approximate measure of the relative importance of crops and livestock, and of different types of crops and livestock in the farm economy. In either case the information can be converted to a percentage and Weaver's combination index may be used.

The method employed is to compare the proportions of an observed distribution with a series of hypothetical distributions to establish the one to which it most closely approximates. Let us take as our example the leading agricultural enterprises in the Nottingham parish of Ruddington, as defined by their total 'man-day' requirements in 1966 (Table 1.10a).

TABLE 1.10a
Leading agricultural enterprises in the Nottinghamshire parish of Ruddington in 1966

Enterprises	Man-days %
Cash crops	50
Pig raising	19
Beef raising	15
Dairying	12
Poultry	3
Sheep	1

The problem is whether to classify the parish as a 1-, 2-, 3-, or 4-enterprise economy. Cash crops occupy half the total labour time. If we add pig raising the total is 69 per cent; adding beef raising this becomes 84 per cent. It would seem reasonable, with so much labour time occupied by three enterprises, to place the parish in the cash crops/pigs/beef category. Calculation will show that this is not the case, and clearly demonstrates the necessity for some objective, if arbitrary, method of classification.

The way in which the calculation is made is given in Table 1.10b. We have in our example six enterprises. Supposing total man-days were divided between two categories only, the hypothetical distribution would show 50 per cent in each and zero in the remainder. Similarly even distribution between all categories gives 17 per cent in each (the approximation is close enough to ignore decimals). Table 1.10b shows the hypothetical distribution expected if the man-days are concentrated in one category only, or are *evenly distributed* between 2, 3, 4, 5, or 6 categories. Weaver's index is a method of determining to which of these hypothetical distributions the observed distribution most nearly approximates.

The method is to obtain the sum of the squares of the difference between each category of the hypothetical and the observed distri-

TABLE 1.10b

	1 Enterprise						d^2
Hypothetical %	100	0	0	0	0	0	
Observed %	50	19	15	12	3	1	
d	50	19	15	12	3	1	
d^2	2500	361	225	144	9	1	3240

	2 Enterprise						
Hypothetical %	50	50	0	0	0	0	
Observed %	50	19	15	12	3	1	
d	0	31	15	12	3	1	
d^2	0	961	225	144	9	1	1340

	3 Enterprise						
Hypothetical %	33	33	33	0	0	0	
Observed %	50	19	15	12	3	1	
d	17	14	18	12	3	1	
d^2	289	196	324	144	9	1	963

	4 Enterprise						
Hypothetical %	25	25	25	25	0	0	
Observed %	50	19	15	12	3	1	
d	25	6	10	13	3	1	
d^2	625	36	100	169	9	1	940

	5 Enterprise						
Hypothetical %	20	20	20	20	20	0	
Observed %	50	19	15	12	3	1	
d	30	1	5	8	17	1	
d^2	900	1	25	64	289	1	1280

	6 Enterprise						
Hypothetical %	17	17	17	17	17	17	
Observed %	50	19	15	12	3	1	
d	33	2	2	5	14	16	
d^2	1089	4	4	25	196	256	1574

bution. The most accurate approximation is shown by the smallest sum of the squares (Table 1.10b), because if the observed percentages are identical with the hypothetical distribution the difference between them would be 0, and the squares of the differences also 0. The 'best fit' therefore is when the sum of the squares of the differences most nearly approaches 0, in our example Case 4, and we are

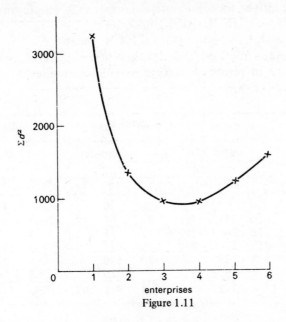

enterprises

Figure 1.11

able to classify the parish as approximating most nearly to a four-enterprise economy based on cash crops/pigs/beef/dairying. The values of the sum of the squares are plotted graphically in Figure 1.11. The closeness of the sum of the squares, and therefore of the goodness of fit, between a three- and a four-enterprise economy, clearly demonstrates the need for an objective method of classification.

While the classification of a single area in this way is of no value, the method is useful when applied to a large number of areas, especially if the information is to be mapped (see Coppock's *Agricultural Atlas of England and Wales*, 1964). It can also be used in any situation that demands classification using a number of variables, provided data are in the form of proportions which add up to 100 per cent. A notable use was by D. Smith (1969) who used the technique to map 'one-industry' and 'two-industry' towns in Lancashire and Cheshire.

Consolidation

5. Numbers employed in manufacturing industry in the U.K. and in N. Ireland by SIC categories (1958) III to XVI (thousands) for 1966.

| N. Ireland | 31 | 3 | 1 | 30 | 14 | 8 | 3 | 55 | 0 | 27 | 4 | 5 | 6 | 4 |
| U.K. | | 871 | 532 | 628 | 2391 | 219 | 858 | 604 | 818 | 60 | 558 | 369 | 322 | 654 | 352 |

(Source: *Annual Abstract of Statistics*, 1970)

Draw Lorenz curves to compare specialization of employment in N. Ireland with that of the United Kingdom.

Calculate the index of concentration for N. Ireland.

6. Using Weaver's method classify the following three Nottinghamshire parishes in terms of percentage man-days allotted to the leading cash crops grown.

Leading Crops (excl. horticulture)	Clipstone			Willoughby			Ruddington		
	Acge.	Man-days No.	%	Acge.	Man-days No.	%	Acge.	Man-days No.	%
Wheat	29	72	1	20	50	5	397	992	13
Barley	926	2315	30	33	82	9	1100	3750	49
Potatoes	78	1422	18	1	18	2	23	416	5
Sugar beet	216	2700	35	6	75	8	130	1625	21
Hay and silage	197	295	4	420	630	70	245	367	5
Others (fodder)	278	908	12	14	56	6	95	561	7
Total	1724	7712	100	494	911	100	1990	7711	100

1.8. THE TRANSFORMATION AND COMBINATION OF DATA

It is often useful to be able to compare or combine scores on different variables and in different units. Consider the data in Table 1.11,

TABLE 1.11

Social and economic characteristics of major countries

	Infant mortality (per 1000 live births)	Food consumption (daily p.c. calorie intake)	Energy consumption (p.c. kg. coal equivalent)
India	139	1940	98
U.S.S.R.	24	3180	4 200
U.S.A.	19	3290	10 774
Indonesia	125	1750	58
Pakistan	142	2260	93
Japan	13	2450	2 828
Brazil	90*	2540	481
Nigeria	140*	2170	29
W. Germany	24	2990	4 850
U.K.	18	3180	5 139
Mexico	69	2620	1 044

Data are for late 1960s from U.N. statistical year books
* = Estimates; no data given

which are measures of economic development and social well-being for the eleven most populous countries in the world, excluding China (for which data were not available).

Suppose we need to compare scores from different columns either individually (e.g. Japan's standing in terms of infant mortality with her standing in terms of energy consumption) or in aggregates (e.g. how does Japan's over-all socio-economic well-being compare with that of West Germany?). In either case, it is usually necessary to transform the raw data so that they are on a common scale of values. Three frequently used transformations will be described, the results of which are set out in Table 1.12 (incomplete to give the reader the opportunity to work out missing values).

TABLE 1.12
Transformed scores for data in Table 1.11

	Ranked				Scaled				Standardized			
	I.M.	F.C.	E.C.	C.	I.M.	F.C.	E.C.	C.	I.M.	F.C.	E.C.	C.
	1	2	3	4	5	6	7	8	9	10	11	12
India	9	8	8	25	2	12	1	15	−1·24	−1·27	−0·81	−3·32
U.S.S.R.	4·5	2·5	4	11	91	93	39	223	−0·92	+1·20	+0·47	+2·59
U.S.A.	3	1	1	5	95	100	100	295	+1·04	+1·42	+2·52	+4·96
Indonesia	8	11	10	29	13	0	0	13	−0·98	−1·65	−0·82	−3·45
Pakistan	11	8	9	28	0	33	1	34	−1·30	−0·64	−0·81	−2·75
Japan	1	7	5	13	100	45	26	171	+1·13	−0·25	−0·04	+0·84
Brazil	7	6	7	20	40	51	4	95	−0·32	−0·07	−0·69	−1·08

(Note: To calculate z-scores use the following:

	Mean	Standard deviation
Infant mortality	73	53
Food consumption	2579	502
Energy consumption	2690	3210

1.8a. Ranking
This is the simplest method, and is perhaps most appropriate where data are unreliable, since it involves discarding some of the possibly inaccurate information they contain. Each column of raw data is ranked separately, as in the first three columns of Table 1.12. It is now possible to make meaningful comparisons between individual figures (e.g. Japan is first in terms of infant mortality, but fifth in energy consumption). It is also possible to produce comparable aggregates by simply adding ranks scored by a country in each variable (thereby obtaining what is sometimes known as Kendall's ranking coefficient), giving a composite score for each country's status in terms of the variables included (Column 4 in Table 1.12).

Two points about the ranking must be made:

(i) Whether one ranks from high to low or vice versa depends upon the object of the comparison being made. In the example given, it is likely that comparisons will be made in terms of well-being, so that each column is ranked from 'best' to 'worst'. Thus, low scores in infant mortality and high scores in food consumption are ranked high. (This is in contrast to correlation procedures described in Chapter 7.) Value judgements are involved here, about which more will be said at the end of this section.

(ii) Where raw scores are tied, an average rank is allotted to each. Thus, U.K. and U.S.S.R. both have a food consumption value of 3180, which means they occupy ranks 2 and 3 on that variable; each is therefore given the rank of 2·5.

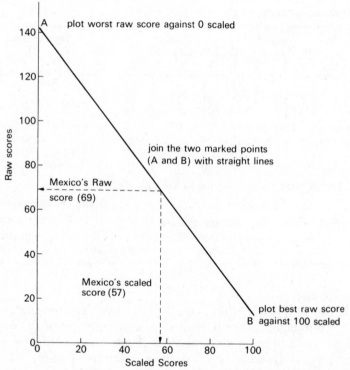

Figure 1.12. Conversion graph for scaling infant mortality data in Table 1.11.

1.8b. Scaling

While ranking involves reducing raw interval to transformed ordinal data, with consequent loss of information (see Introduction, p.xvii), scaling maintains the intervals between individual values in their correct proportions. The 'best' score in each column is made equal to 100, the worst to 0 (or some other arbitrary round figures), and intermediate values are determined by graph (Figure 1.12) or by the following formula:

$$S = \frac{R - R_{worst}}{R_{best} - R_{worst}} \times 100 \qquad \textit{Formula 1.8}$$

where S is the scaled and R the raw value. (A similarity between this and Formula 1.6 for the Lorenz curve index of concentration will be noticed—why?) Thus, the scaled value for U.K. infant mortality is:

$$S = \frac{18 - 142}{13 - 142} \times 100 = \frac{-124}{-129} \times 100 = 96.$$

One can use scaled values in the same way as ranks, either to compare individual scores or to produce composite scores (Columns 5 to 8 in Table 1.12).

1.8c. Standardization
While scaling makes data comparable by equalizing the extreme range of scores on each variable, standardization does so by equalizing the average range, as expressed by the standard deviation.

The standardized score is otherwise called a z-score (pronounced zee-score), and is obtained from the formula:

$$z = \frac{R - \bar{R}}{\sigma} \qquad \textit{Formula 1.9}$$

where R is an individual raw score on a given variable, and $\bar{R}$ and o are respectively the mean and standard deviation of all scores on that variable. Thus, the z-score for Mexico's energy consumption is:

$$z = \frac{1044 - 2690}{3210} = -0.51$$

meaning that Mexico's per capita energy consumption is just over half a standard deviation below the mean for the eleven countries listed. The effect of standardization is to give varying weight to extremes, as can be seen by comparing the extreme range of standardized scores for infant mortality and energy consumption (from $+1.13$ to -1.30, and from $+2.52$ to -0.82 respectively—Columns 9 and 11 in Table 1.12).

Again, if one is comparing data on a best–worst scale, then values that are better than average should be given positive z-scores, and those worse should have negative z-scores. This will entail reversing the signs of all z-scores in some variables—in the example given, all infant mortality z-scores have had their signs reversed. They can now be used to make direct comparisons of individual scores (e.g. U.S.A. has an energy consumption which is two-and-a-half standard deviation above the mean, but an infant mortality rate only one standard deviation above the mean), or they can be aggregated to produce another composite score of socio-economic development (Column 12 in Table 1.12).

1.8d. Logarithmic transformation

A further frequently used transformation has a different purpose. Whereas the three just described bring variables using different units of measurement on to a common scale, so that like can be compared with like, a logarithmic transformation may be used to modify the interval between scores on the same scale. It has the effect of reducing the difference between high scores in relation to the difference between low scores, as the following example shows:

Raw Score (R)	Interval	Log R	Interval
1000		3	
	900		1
100		2	
	90		1
10		1	
	9		1
1		0	

This is because the difference between the logs of two numbers expresses the *proportional* difference between the numbers—the proportional difference between 1000 and 100 is the same as the proportional difference between 10 and 1. (Strictly, one should specify the base' of a log with a suffix—thus, 'the log of 100 to base 10 is 2' should be written: $\log_{10} 100 = 2$, which is the reverse of saying that $10^2 = 100$. Logs may be to any base—e.g. $\log_5 125 = 3$, meaning $5^3 = 125$. But in this book, all logs are taken to be base 10, which are the ones most commonly found in log tables.) Of the variables in Table 1.11, there are good grounds for transforming energy consumption into logs (and then scaling or standardizing the logs). Reasons include

(i) Energy consumption tends to grow proportionally (as does GNP—growth rates are generally referred to in percentage terms) so that the difference between countries might also be expressed in proportional terms.

(ii) Very high energy consumption is particularly susceptible to diminishing returns (e.g. through pollution, waste, and resource exhaustion), so that on a best-worst scale the high values accorded to, say, U.S.A., might be scaled down to take some account of the diseconomies they imply.

Table 1.13 presents the energy consumption data doubly transformed—first the raw data are logged, and these are then scaled. Again the last four rows are left for the reader to complete.

TABLE 1.13
Double transformation of energy data from Table 1.11

	Raw	Logs	Scaled Logs
India	98	1·99	21
U.S.S.R.	4 200	3·62	84
U.S.A.	10 774	4·03	100
Indonesia	58	1·76	12
Pakistan	93	1·97	20
Japan	2 828	3·45	77
Brazil	481	2·68	47
Nigeria	29		
W. Germany	4 850		
U.K.	5 139		
Mexico	1 044		

It will be seen that, although the U.S.A. still scores 100 in the scaled column (logs do not alter ranks, and the highest rank must score 100 in scaling), Japan now scores 77 instead of 26 from scaling of the raw data—i.e. the gap between U.S.A. and Japan has been reduced from 74 to 23, which probably accords more closely with the difference in the economic strength of the two countries. (On the other hand, while differences between high values are diminished those between low values are enlarged, e.g. scaled raw values for Indonesia's and Pakistan's energy consumption are 0 and 1 respectively; the scaled logs are 12 and 20. This may not be justified in view of the unrealiability of data from less developed countries.)

1.8e. Value judgements
It will be clear that the data used in this section are value loaded. What is 'best' and 'worst'? In the case of infant mortality we should probably all agree, but what about food and energy consumption? Is more necessarily better? And when it comes to producing composite scores of social well-being, economic health, or some other macro-statistic, more questions arise: are we aggregating the most appropriate variables? (For example, should literacy, crime-rates,

incidence of nervous breakdowns, be among the ingredients of social well-being? There is a considerable body of literature on social indicators of this sort.) Are the data sufficiently reliable to justify scaling or standardization rather than ranking?

Should any of the data be logged? Should all variables be weighted equally, or should infant mortality, for instance, count for more than, say, energy consumption? And if so, by how much? The effect of this last question—weighting—is illustrated by comparing composite scores for West Germany and Japan, derived from scaled Infant mortality and Energy consumption alone, (*a*) by assigning equal weight to both variables, and (*b*) by weighting Infant mortality and Energy consumption in the ratio of 3 to 1.

		Scaled scores				
	Equal weighting			*Unequal weighting*		
	Infant mortality	*Energy consumption*	*Com-posite*	*Infant mortality X 3*	*Energy consumption*	*Com-posite*
W. Germany	91	45	136	273	45	318
Japan	100	26	126	300	26	326

When the two variables are given equal weight, West Germany has the higher (better?) score; when infant mortality is weighted 3 times energy consumption, Japan has a higher score.

The above questions are all rhetorical—there are no definitive answers, other than that judgements about methods must be made in the light of objectives—e.g. if one is assessing social well-being, infant mortality might be more heavily weighted than energy consumption; if one is assessing economic performance, the reverse might be more appropriate. Geographers cannot escape making value judgements, even when they are quantifying. It is essential that we recognize the limits of our objectivity, and the extent to which values enter into our work.

Consolidation

7. Complete Table 1.12. Rank the composite scores in Columns 4, 8 and 12. Comment on the differences and similarities in rank between the three columns.

8. Complete Table 1.13. Substitute the last column of this table for Column 7 in Table 1.12 and, by adding scores in this column to those from Columns 5 and 6, calculate a new composite score based

on scaled values. Are there any differences between the ranks of the scores thus obtained and those previously obtained in Column 8 of Table 1.12?

9. Weight the ranked scores in Table 1.12 (Columns 1 to 3) in the ratio of 3 (Infant mortality) to 2 (Food consumption) to 1 (Energy consumption). Add the results to obtain a weighted ranking coefficient, and compare with the unweighted results already obtained in Column 4.

MEASURES OF SPATIAL DISTRIBUTION

2.1. INTRODUCTION

THE previous chapter outlined various ways of summarzing and comparing what might be termed 'place data', in which each set of raw data relates to the same place, so that location is a constant. The mean annual rainfall of Sydney and the Gibbs-Martin index of diversification for the export trade of the United Kingdom (pp.11 and 31) summarize sets of raw data relating to single places—Sydney and the United Kingdom respectively. This chapter will consider the summary of 'space data'—i.e. data in which location is a variable, and which can therefore be expressed as spatial distributions.

The usual way of depicting spatial distributions is by maps, and it is a commonplace that the map is the basic tool of the geographer. But maps by themselves are mute, and it is a major task of the geographer to describe, compare, and interpret them; to make explicit what is implicit in them; to analyse them. Statistics are usually helpful, and sometimes essential in this process.

Broadly, there are two approaches to map analysis—that of the 'pen-portrait', and that of the 'statistical profile'. In the former, patterns are identified intuitively, and characterized with whatever verbal facility one can command. In the latter, one tries to find objective measures of various facets of the mapped distributions (such as density, degree of clustering, etc.), and present these rather like a set of vital statistics or the marks on a school report. The two approaches are complementary rather than mutually exclusive, and the fact that we are here concerned with the statistical approach does not imply any claim for its superiority. Statistics by no means say all there is to say about distribution patterns, but they have three special attributes:

   (i) They add precision to qualitative verbal description.
   (ii) By offering objective, quantitative criteria, they facilitate the making of comparisons between distributions.
   (iii) They may draw attention to characteristics unlikely to be noticed by intuitive inspection.

2.2. FOUR TYPES OF SPATIAL DATA

Methods by which maps can be statistically summarized vary with the locational characteristics of the data being mapped. Techniques for four types of spatial data will be discussed in this chapter:

   (i) Point distributions—in which each item is allocated to a discrete point on the map, as in the case of a dot map.

   (ii) Line distributions—in which each item is represented by a straight, curved, or irregular line, as on a map showing a drainage system or road network.

   (iii) Discrete areal distributions—in which each item or statistic refers to a distinct area of the map. Choropleth maps are of this kind.[1]

   (iv) Continuous areal distributions—in which data relate to sample points in a spatial continuum, and which are generally mapped by isopleths—e.g. maps of temperature and rainfall.

2.3. POINT DISTRIBUTIONS

Three kinds of measure will be considered: (a) central location; (b) dispersion and clustering; (c) distance from a point.

2.3a. Central location

The central location of a point distribution can be determined by methods analogous to the measures of central tendency for a numerical distribution (p.4), but results are generally expressed as locations on a map rather than by numbers. *The median centre* is the point of intersection of two perpendicular median lines (usually north-south and east-west), each of which bisects the distribution of points (lines AB and CD on Figure 2.1). Its precise location will vary with the

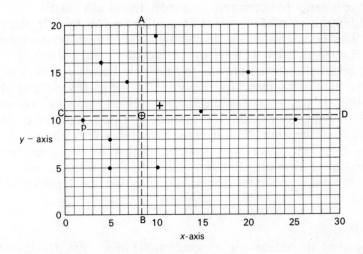

Figure 2.1. The mean and median centres of a hypothetical point distribution

orientation of the perpendicular lines, but the variation will be small if there is a large number of points. The arbitrary effect of the orientation of the median lines can be minimized by drawing a second pair at 45° to the first, and then taking the mid-point between the intersections of the two pairs of median lines as the median centre.

The _mean centre_ can be regarded as the 'centre of gravity' of a distribution, since each item is weighted according to its location. It is determined by:

(i) Arbitrarily drawing an x- and y-axis, generally to the south and west of the distribution of points;

(ii) Determining the x- and y-coordinate of each point, e.g. in Figure 2.1, p has the coordinates $x = 2, y = 10$.

(iii) Calculating the mean of the x- and y-coordinates, as follows:

$$\bar{x}_c = \frac{\sum x_c}{n_x}$$

$$x\text{-coordinates:} \quad \frac{2 + 4 + 5 + 5 + 7 + 10 + 10 + 15 + 20 + 25}{10}$$

$$= \frac{103}{10} = 10 \cdot 3$$

$$y\text{-coordinates:} \quad \frac{10 + 16 + 5 + 8 + 14 + 5 + 19 + 11 + 15 + 10}{10}$$

$$\bar{y}_c = \frac{\sum y_c}{n_y}$$

$$= \frac{113}{10} = 11 \cdot 3.$$

The median and mean centres in Figure 2.1 are indicated by O and + respectively. They do not, as a rule, differ greatly in location, and as the former is easier to determine, it is more frequently used.

Some interesting uses are made of median centres by J. E. Martin (1966). This book contains a chapter by C. M. Brown on the industries of London's New Towns, which includes a map showing the place of origin of factories moving to four New Towns. Figure 2.2a is abstracted from this map, and Figure 2.2b shows (by dots at the intersection of the lines) the median centres of places of origin of factories moving to Basildon, Bracknell, Crawley, and Stevenage. It shows that the four New Towns have tended to attract more factories from the quarter of London closest to them, implying some kind of gravitational link between old and new locations. More than eight miles separates the median centres of Basildon and Bracknell.

2.3b. Dispersion

The degree of dispersion (or concentration) of a point distribution may be measured in relation to any one of three criteria:

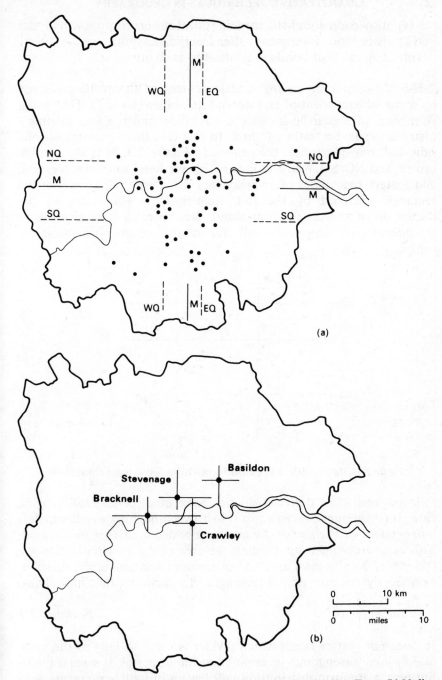

(a)

(b)

Figure 2.2. (a) Origins of factories moving from London to Crawley New Town; (b) Median Centres and interquartile distances for origins of factories moving from London to four New Towns.

(i) dispersion about the median (or mean) centre;
(ii) dispersion about some other specific location;
(iii) dispersion of points in relation to each other.

2.3bi. Dispersion about the median centre.   This can be measured
by a spatial equivalent of an inter-quartile range (see p.7). The spatial
equivalent of a quartile is called a quartilide, and is a line dividing a
distribution in the ratio of three to one (i.e. three-quarters on one
side, and one-quarter on the other). In Figure 2.3, M is the median
centre, and NQ, SQ, WQ, and EQ are the northern, southern, western,
and eastern quartilides of an imaginary distribution. The area of the
rectangle enclosed by the four quartilides is a measure of the
dispersion of the distribution about the centre: a large area means
considerable dispersion; a small area means concentration near the
centre.

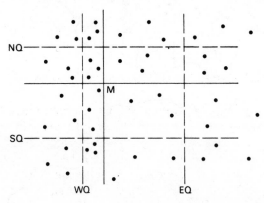

Figure 2.3. Median centre and quartilides of a hypothetical point distribution

It is possible to derive a simple index of dispersal (*Id*), ranging
from 0 (maximum concentration) to 1 (maximum dispersal) which is
independent of the size of the area over which points are distributed.
This can be obtained by dividing the area of the quartilide rectangle
(*Q*) either by the total area (*A*) of the unit containing the distribution; or by the area of the rectangle (*A′*) enclosing that unit. Thus,

$$Id = \frac{Q}{A} \text{ or } \frac{Q}{A'} \qquad\qquad \textit{Formula 2.1}$$

It does not matter much which divider is used, so long as the same
one is used throughout in comparing the dispersal of several distributions. A uniform distribution will have a quartilide rectangle with

an area about a quarter that of the whole distribution (Figure 2.4a), and will have an index of about 0·25. A peripheral distribution like that in Figure 2.4b (which is not unlike the distribution of population in Australia) will have an index close to 1, indicating maximum dispersal about the median centre. A very clustered distribution will have an index approaching zero.

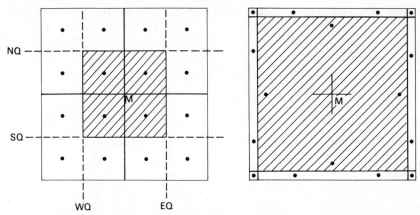

Figure 2.4. (a) and (b) Quartilide rectangles for uniform and peripheral point distributions

The lengths of the lines through the median centres in Figure 2.2b indicate the dimensions of the quartilide rectangles for the distributions there summarized. Thus, the quartilide rectangle for Basildon (on the map) has dimensions of 0·4 in. and 0·4 in., giving an area of 0·16 square in. The area of Greater London on the map, using the same units, is difficult to calculate, but the area of the rectangle enclosing the whole of Greater London has dimensions of 3·7 in. and 3·35 in., giving an area of 12·4 square in. An index of dispersion for Basildon can therefore be calculated:

$$Id = \frac{Q}{A'} = \frac{0·16}{12·4} = 0·013$$

This figure, by itself, is not very meaningful. But if similar figures are calculated for the other New Towns, one is able to compare the degree of dispersion of the places of origin of factories moving out of London into the New Towns.

Consolidation
1. Calculate indexes of dispersion for factories moving to Crawley, Bracknell, and Stevenage.
   (The results obtained could provide a starting-point for further investigation).

2.3bii. Dispersion about any location. The quartilide rectangle measures dispersion about the median centre. An alternative measure, which describes dispersion about any chosen location (including the median centre if required) can be obtained by grouping points in terms of their distance from the selected location. This is particularly appropriate when examining the distribution of different kinds of shops in relation to a city centre (which may be far removed from

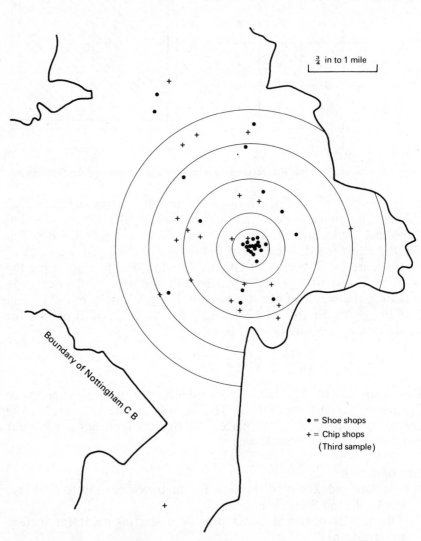

Figure 2.5. The distribution of shoe shops and chip shops in Nottingham C.B.

any median centre), and an example will be given to demonstrate how this measure may be obtained and used.

Figure 2.5 shows the distribution of two kinds of retail establishment in a large city. Only those shops listed in the classified telephone directory were plotted, and in the case of fish and chip shops, a one in three sample was taken—every third name in the directory. The circles on the map are centred on the city centre, and have radii of $\frac{1}{4}$, $\frac{1}{2}$, $1\frac{1}{2}$, and 2 miles.

The two distributions are clearly very different, and well illustrate the effect of differing 'ranges' of central place functions upon their distribution patterns. (The 'range' of a central place function is the distance that people are prepared to travel to obtain it.) A contrast as sharp as this hardly needs statistical support, but quantitative techniques would be useful in less extreme cases, or if one were comparing a number of such distributions, with a view to ranking them in order of centrality.

The visual contast can be made more explicit by drawing cumulative frequency curves of the number of shops within each of the circles on the map. This has been done in Figure 2.6, and shows the following: (i) in both types of shop, between 80 and 90 per cent occur within two miles of the city centre; (ii) over 50 per cent of the shoe shops occur within $\frac{1}{4}$ mile of the city centre, while another 40 per cent occur beyond $\frac{1}{2}$ mile from the centre, so tht there are two distinct elements in the distribution pattern of shoe shops—one element concentrated, and the other dispersed; (iii) the fish and chip shops are much more evenly distributed.

It should be noted, however, that a uniform distribution of shops would not be represented by a straight-line graph, since the number of shops within a circle would be proportional to its area, and therefore to the *square* of its radius. For comparison, a curve has been added to Figure 2.6 to represent a distribution which, like the shoe and chip shops, has about 80 per cent of its locations within 2 miles of the city centre; and which, unlike the shoe and chip shops, is uniformly spread within the two-mile circle.[2]

A comparison of gradients will now show more precisely at what distances from the centre the main concentrations of shops occur. For instance, the gradients of both shoe-shop and chip-shop curves are steeper than the uniform curve between $\frac{1}{2}$ mile and 1 mile from the centre, indicating some concentration in this ring.

Finally, a statistical measure of dispersion can be read from the graph: the *median distance*, being the radius of a circle enclosing just 50 per cent of the distribution. In the case of shoe shops, it is just under $\frac{1}{4}$ mile; for chip shops, it is just over 1 mile; and for a uniform

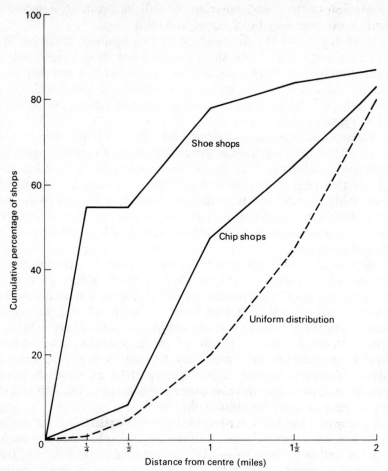

Figure 2.6. Cumulative frequency curves to show the proportion of shoe shops and chip shops within different distances of the city centre of Nottingham.

distribution (assuming 80 per cent of the distribution to be within 2 miles of the centre), the median distance is just over $1\frac{1}{2}$ miles. Both shoe shops and chip shops are seen to be 'centrally oriented', but in different degrees.

2.3biii. Dispersion of points in relation to each other   Quartilide rectangles and median distances are measures of dispersion (or clustering) about specified points. It is possible, however, for a distribution to be both dispersed (in relation to some central point) and also to be highly clustered (in that individual locations occur in groups). Clustering can occur anywhere, and not necessarily near a central point—e.g. the population of Australia, half of which is concentrated in five peripheral metropolitan clusters.

Clustering of points in relation to each other is a function of the degree of uniformity of their distribution, and can be measured on a continuum, ranging from maximum dispersal (in a uniform, triangular distribution), through random clustering and dispersion (the kind of irregular distribution that could be due to chance), to maximum concentration. There is a diagram on p.269 which shows these three landmarks along the continuum.[3]

Methods of measuring the degree of clustering of points in relation to each other are of two kinds:

(a)   Those which measure the distance of a specified number of nearest neighbours from each point. The most widely used of these is the nearest neighbour index, but as this has inferential properties, its discussion is deferred until Chapter Nine (see p.270).

(b)   Those which count the number of neighbours within a specified distance of each point.

The mean number of neighbours within a specified distance
It has been said that clustering (and therefore the distance between neighbours) is a function of the degree of uniformity of a distribution. But it is also a function of the number (and therefore the density) of points in the distribution. If one wishes to have an index which reflects the degree of uniformity of the distribution alone, it must be made independent of density. This can be done by making use of the fact that the *mean* distance $(\bar{D})$ of a point from its nearest neighbour in a *random* distribution is given by the formula:

$$\bar{D}_{ran} = \frac{1}{2\sqrt{\Delta}} = \frac{1}{2 \times \frac{N}{A}} , \qquad\qquad Formula\ 2.2$$

where Δ is the density of points—i.e. the number of points (N) divided by the area (A) under consideration. It follows that, in a random point distribution, of whatever density, there will, on average, be one neighbour within the distance $(\bar{D}_{ran})$ of each point—i.e. the mean number of neighbours per point within this distance is 1. For uniform and clustered distributions, the corresponding means are less and more than 1 respectively.

The formula is therefore used to determine the 'specified distance' and thereby the degree to which a distribution deviates from random. When applied to the shoe shops within 2 miles of the city centre in Figure 2.5, and assuming there are no shoe shops within this radius outside the city boundary, then:—

number of points (N) = 27,

area = $\pi \times 2^2$ = 12·5 square miles,

$$\bar{D}_{ran} = \frac{1}{2\sqrt{\dfrac{27}{12\cdot5}}} = 0\cdot34 \text{ miles}$$

As the map in question is on a scale of $\frac{3}{4}$ inch to 1 mile, the easiest way to determine the number of neighbours within 0·34 miles of each shoe shop is to draw a circle 0·34 $\times \frac{3}{4}$ = 0·26 inches in radius on a piece of tracing-paper with the centre clearly marked, place it over each point in turn, and count the number of neighbours within the circle each time. The following results are obtained:

No. of points with no neighbours within the circle: 6
No. of points with 1 neighbour within the circle: 4
No. of points with between 2 and 15 neighbours: 0
No. of points with 16 neighbours within the circle: 17

These figures can be tabulated as follows:

No. of neighbours (n)	0	1	2	3	...	15	16
Frequency (f)	6	4	0	0	...	0	17

The mean number of neighbours within the circle is given by $\Sigma fn/\Sigma n$, i.e.

$$\frac{6\times0 + 4\times1 + 17\times16}{6 + 4 + 17} = \frac{276}{27} \simeq 10.$$

Had the distribution been random, the average number of neighbours within the circle would have been about 1; a figure greater than 1 indicates a tendency towards clustering; a figure as high as 10 indicates very considerable, and no doubt purposeful clustering—in this case, to gain from the external economies of a city-centre location, in all probability. This result is fairly obvious from a casual glance at the map. Let us therefore look at part of the distribution, where the result is less predictable. If we examine the shoe shops outside the half-mile circle around the city centre, and within the two-mile circle, using the same method, we find that:

(i) The specified distance D_{ran} = 0·59 miles, which on the map is 0·59 $\times \frac{3}{4}$ = 0·44 inches.
(ii) The frequency (f) of shops with n neighbours is:

n	0	1	2	more than 2
f	2	4	4	0

(iii) The mean number of neighbours per shop is:

$$\frac{0 \times 2 + 1 \times 4 + 2 \times 4}{2 + 4 + 4} = \frac{12}{10} = 1\cdot2.$$

This is close to random, with a slight tendency towards clustering.

Consolidation
2. Calculate the number-of-neighbours index for the chip-shop distribution shown on Figure 2.5 within 2 miles of the City centre. (The fact that this is a one in three sample will not materially affect the result, as the index is independent of the density of points.)

2.4. LINE DISTRIBUTIONS–NETWORKS
The lines on maps (other than isopleths and grid lines, which do not represent actual features) generally depict either channels of movement (rivers, routeways) or boundaries (which are often barriers to movement). Where two or more lines meet, a *node* or *vertex* is formed; the lines joining nodes are called *links, arcs,* or *edges.* A system of nodes and links is called a network, and the analysis of networks has developed into a major field of study, with its own jargon (of which the terms here italicized are examples). Networks may be *planar* (two-dimensional) or *non-planar* (three-dimensional). Geographers are principally interested in the medium-scale features of the earth's surface; horizontal dimensions are generally far greater than vertical, and edges usually cross in the same horizontal plane; therefore the networks with which geographers are principally concerned are planar, or at least can be regarded as planar. Exceptions include air routes, underground railway systems, road systems with flyovers and subterranean drainage channels. Non-planar networks which are not generally studied by geographers include electric circuits and the blood circulatory system. Limitations of space forbid a comprehensive survey of networks, and we shall confine ourselves to a consideration of transport and boundary networks.

2.4a. *Transport networks*
These can be analysed from three points of view:
   (i) the accessibility of individual nodes;
   (ii) the over-all characteristics of a route system;
   (iii) the over-all characteristics of traffic flows.

2.4ai. The accessibility of a node can be expressed in terms of either the links (routes) or the traffic flow by which it is connected with the rest of the network. Thus, the 'nodality' of towns can be simply measured and compared by counting the number of roads or the frequency of trains converging upon each; that of shopping centres by the number of bus services or the frequency of buses passing through them. It should be stressed that accessibility is a relative concept, and can only be measured in relation to specified locations and modes of transport.

A method only slightly less straightforward involves the construction of an accessibility matrix. In Figure 2.7, A, B, C, D, and E are five nodes joined by a network of links. The numbers represent the lengths of the links in miles (or they could be taken to represent the time or cost needed to travel along them). Accessibility should be measured in terms of the needs of potential travellers. If the first priority is to minimize the number of changes or transhipments that have to be made, and if each node represents a transhipment or break-of-bulk location (e.g. a railway terminus), then one should set up a matrix showing the minimum number of changes that must be made between nodes, as in Table 2.1.

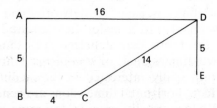

Figure 2.7. A hypothetical network of five nodes and five links

TABLE 2.1
Matrix showing changes necessary between nodes A to E

	A	*B*	*C*	*D*	*E*	*Total*	*Rank*
A	–	0	1	0	1	2	2
B	0	–	0	1	2	3	4
C	1	0	–	0	1	2	2
D	0	1	0	–	0	1	1
E	1	2	1	0	–	4	5

(The last two columns are grouped under the heading *Accessibility*.)

D is seen to be the most accessible node of the network, *in terms of the specified criteria*—i.e. in relation to the other nodes, which are all given equal weighting (implying that the same number of people will want to travel to and from each); excluding routes other than those

represented by the edges in the diagram; and in terms of the number
of changes required.

If any one of these criteria is changed, a different ranking of
accessibility may result. For instance, one may be interested in
reducing to a minimum the mileage rather than the number of
changes. In that case, a mileage matrix is required (Table 2.2). C is
seen to be the location from which journeys can be made to the
other nodes in the network with a minimum aggregate mileage.

TABLE 2.2
Mileage matrix between nodes A to E

	A	B	C	D	E	Accessibility Total	Accessibility Rank
A	0	5	9	16	21	51	3
B	5	0	4	18	23	50	2
C	9	4	0	14	19	46	1
D	16	18	14	0	5	53	4
E	21	23	19	5	0	68	5

The question of weighting has been mentioned. Suppose the nodes
A, B, C, D, and E represent towns with populations of 10 000,
20 000, 30 000, 40 000, and 50 000; and suppose that a new factory
(e.g. a bakery or a dairy) is to be built to serve these five towns.
Assuming that the number of deliveries required to the five towns is
proportional to their populations, and that the object is to minimize
aggregate delivery mileage, then the mileage matrix must be weighted
in the ratio of 1 to 2 to 3 to 4 to 5, as in Table 2.3:

TABLE 2.3
A weighted mileage matrix

	A	B	C	D	E	Accessibility Total	Accessibility Rank
A	0 X 1	5 X 2	9 X 3	16 X 4	21 X 5	206	5
B	5 X 1	0 X 2	4 X 3	18 X 4	23 X 5	204	4
C	9 X 1	4 X 2	0 X 3	14 X 4	19 X 5	168	3
D	16 X 1	18 X 2	14 X 3	0 X 4	5 X 5	119	1
E	21 X 1	23 X 2	19 X 3	5 X 4	0 X 5	144	2

D is the location with the lowest aggregate delivery costs, and, other things being equal, the factory should be located there.

Finally, let us turn our five nodes back into transhipment points (e.g. railway junctions) of equal weighting, in order to show how two matrices can be combined. No traveller or consigner of goods is exclusively interested in minimizing mileage, or in minimizing the number of transhipments he must make. He will take both factors into accounts. This can be done if some equivalence can be established between transhipments and mileages. Let us suppose that the cost and inconvenience of a transhipment (or change of trains) is the equivalent of an additional ten miles of travel. Then the results of the first and second matrices set out above can be combined as in Table 2.4:

TABLE 2.4

Combination of transhipment and mileage data to obtain gross accessibility index

	Transhipments in equivalent miles (from first matrix) X	*Actual mileage (second matrix)* Y	*Gross accessibility* X + Y	*Rank*
A	2 X 10 = 20	51	71	3
B	3 X 10 = 30	50	80	4
C	2 X 10 = 20	46	66	2
D	1 X 10 = 10	53	63	1
E	4 X 10 = 40	68	108	5

Once again, D is seen to be the most accessible node in the network, but the rank order of the other nodes is changed.

Consolidation

3. Below (Table 2.5) is set out a mileage matrix of road distances separating six of Britain's major cities (rounded to the nearest ten miles). Determine their relative accessibility by road in relation to each other, (a) giving each city equal weight, and (b) weighting them in proportion to the populations of their conurbations (roughly, London 9, Birmingham 3, Manchester 3, Liverpool 2, Leeds 1, Newcastle 1).

TABLE 2.5
Mileage matrix of road distances separating six British cities

	London	Birmingham	Manchester	Liverpool	Leeds	Newcastle
London	0	110	180	200	190	270
Birmingham	110	0	80	90	110	200
Manchester	180	80	0	40	40	130
Liverpool	200	90	40	0	70	150
Leeds	190	110	40	70	0	90
Newcastle	270	200	130	150	90	0

2.4aii. General characteristics of a route network A network of
routes can present a very confusing pattern when mapped. Words
such as 'grid-iron' and 'radial' are useful descriptive analogies for
some networks, but many resemble no such regular stereotypes.
Among statistical measures that may aid, and add precision to,
description, two are discussed in this section: (i) measures of density;
and (ii) a measure of route sinuosity. A third type of measure,
connectivity, is considered in the following section dealing with
traffic flows, but this can also be applied to routeways.

Measures of route density
Route density is usually expressed in miles of routeway per hundred
square miles of territory, or per 10 000 people. The measurement of
routes can, however, be tedious, especially in the case of roads,
which generally constitute a greater mileage and tend to be less
straight than other transport arteries. A useful substitute for
measuring hundreds of miles of roads is to count the number of road
junctions, as this has been shown to correlate well with road mileage
in developed countries. Thus, if one wishes to compare the road
density on two Ordnance Survey map sheets, a count of junctions in
a random sample of, say, fifty grid squares on each map would form
a useful measure. (For the method of obtaining a random sample,
and of assessing the reliability of comparisons based on samples, see
Chapters 5 and 6 respectively.)

A measure of route sinuosity
The most direct route between two points on the earth's surface is
along the arc of a great circle (if we discount boring through the
earth's crust). For short distances, this approximates closely to a
straight line on a map. In practice, it is often not possible to travel in
a direct line between two nodes; detours have to be made to avoid
obstacles or to take in other places on the way, or simply because no
direct route has been constructed.

The *detour index* (D.I.) measures the amount of detour of the shortest route connecting two nodes:

$$\text{D.I.} = \frac{\text{shortest route distance between A and B}}{\text{shortest distance between A and B}} \times 100.$$

Formula 2.3

Thus, the shortest road distance between Inverness and Thurso is 122 miles (*Readers' Digest Book of the Road*), and the straight-line distance, measured from an atlas is 81 miles. Therefore the detour index for the Inverness-Thurso road is $(122/81) \times 100 = 150$. The higher the value of the detour index, the greater the sinuosity of the route.

The detour indexes for all nodes in a network can be set out in a matrix, and both nodal means and a mean detour index for the whole network can then be calculated (Table 2.6):

TABLE 2.6
Detour indexes for road network connecting four towns in N. Scotland

	Thurso	Inverness	Fort William	Ullapool	Nodal mean
Thurso	—	$\dfrac{122 \times 100}{81} = 150$	135	159	148*
Inverness	150	—	118	129	133
Fort William	135	118	—	153	135
Ullapool	159	129	153	—	147

*Nodal mean $= (150 + 135 + 159)/3 = 148$

The mean detour index for the whole network is obtained by averaging the four nodal means: $(148 + 133 + 135 + 147)/4 = 141$.

Consolidation
4. Figure 2.8 contains three road networks in different parts of Britain, each network connecting four towns. The *Scottish* network has already been discussed. The numbers in the map matrix for *Wales* indicate route mileage (numerator) and straight-line distance (denominator). The numbers in the matrix for *East Anglia* are the detour indexes of the roads connecting the four towns shown.
    (a)   Calculate detour indexes for the road connections between each of the four Welsh towns.
    (b)   Draw up a detour index matrix for the Welsh network, and calculate the Mean Detour Index of this and the East Anglian networks. Compare the results with those given for northern Scotland, and suggest reasons for the differences.

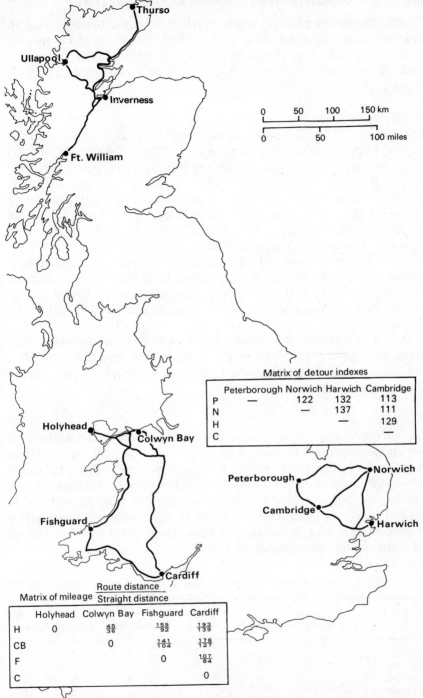

Figure 2.8. Road networks connecting four towns in each of three parts of Britain

2.4aiii. Measurements of overall traffic flow   A transport network can be thought of either as a system of routeways, or as a system of traffic flow; either as channels of movement, or as the movement itself. Two measures of traffic flow will now be outlined: (i) traffic density; (ii) connectivity.

Traffic density

The amount of traffic using a network will, in part, depend upon the size of population and the area served by the network, and also on the length of routeway. Therefore, meaningful comparisons between traffic flow on different networks are made in terms of traffic density, which may be expressed by the amount of traffic divided by either (i) the population or area served by the network, or (ii) the length of the routeway. Furthermore, the amount of traffic is generally expressed in one of two ways:

(a)   The number of vehicles using the network in a given time period—such a figure may be impossible to obtain (e.g. the number of cars on a road network), but a substitute can be found by calculating the mean of a series of traffic counts taken at selected points in the network.

(b)   The number of vehicle-miles travelled on the network in a given time period (a vehicle-mile is one vehicle travelling one mile; ten vehicle-miles is one vehicle travelling ten miles, or ten vehicles travelling one mile each).

Connectivity

Connectivity may be defined as the degree to which the nodes of a network are *directly* connected to each other. The concept is particularly useful in describing public transport systems (private transport is usually direct or 'door to door'—that is its chief attraction). Thus, the connectivity of the London Underground with respect to six main line termini may be expressed in the following connectivity matrix (Table 2.7), in which a 1 indicates a direct link (no change necessary), and a 0 indicates no direct link.

TABLE 2.7

Underground connectivity matrix for London main line termini

	Paddington	Euston	Liverpool St.	Waterloo	Victoria	Kings Cross
Paddington	—	0	1	0	1	1
Euston		—	0	1	1	1
Liverpool St.			—	0	1	1
Waterloo				—	1	0
Victoria					—	1
Kings Cross						—

Similar matrices could be constructed for the network of bus services connecting the neighbourhoods of a town; the air lines serving the major cities of a country; or the rail services connecting the towns of a region. It should be noted, in passing, that, in the case of the London Underground, the connectivity matrix describes both the route network and the traffic network, since each route system has a self-contained traffic system (Northern Line, Bakerloo Line, etc.). This is not true of bus services, air services (which have no physical route-ways), or surface train services.

If it is required to compare the connectivity of two or more networks, a connectivity index, which summarizes the matrix, may be useful. Two such measures are given here.

The beta index β: This is simply the number of links or edges in a network (e). divided by the number of nodes (n):

$$\beta = \frac{e}{n} \qquad \qquad \textit{Formula 2.4}$$

In the above example, e (the number of 1s in the matrix) is 10, and n (the number of stations) is 6, therefore, $\beta = 10/6 = 1\cdot67$. The beta index has been widely used, but it has one serious weakness: the range of possible values depends on the number of nodes in the network. A six-node network has a maximum beta value of $2\cdot5$; a nine-node network has a maximum value of 4. Therefore, a beta index should only be used to compare networks with the same number of nodes, which severely limits its usefulness.

The connectivity index: This expresses the number of links in a network as a ratio of the maximum number of links possible. It can be shown that the maximum number of possible links is $\frac{1}{2}n(n-1)$, where n is the number of nodes in the network. Therefore,

$$\text{the connectivity index, } C = \frac{e}{\frac{1}{2}n(n-1)}, \qquad \textit{Formula 2.5}$$

where e is the number of links or edges in the network, and the denominator is the maximum number of links possible. In the London Underground example, in which $e = 10$ and $n = 6$,

$$C = \frac{10}{(\frac{1}{2} \times 6 \times 5)} = \frac{10}{15} = 0\cdot67$$

i.e. connectivity is two-thirds the maximum.

Consolidation

5. Figure 2.9 is a map of a part of the Paris Metro, showing the lines serving five main railway termini. Construct a connectivity matrix similar to that made out for the London Underground, and compare the connectivity of the two systems in relation to the main line termini, using the connectivity index.

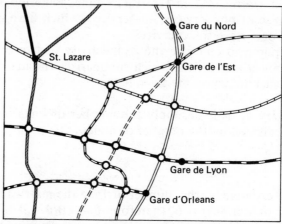

Figure 2.9. Part of the Paris Metro serving five main line stations. **O** denotes stations where passangers may change from one line to another.

2.4b. Boundary networks

The earth's surface is a palimpsest of areal units, ranging from oceans and continents to fields and buildings. They all have boundaries, which may be visible (such as a garden fence) or invisible (such as a parish boundary), and may either be clearly defined or occupy broad zones which can only nominally be represented by lines on a map (e.g. the boundary of an urban sphere of influence). A map which distinguishes areas is necessarily a map containing one or more boundary networks which divide the map into cells of varying size and shape. The measurement of *size* is straightforward and need not detain us long. If a cell has an irregular shape, the simplest way of estimating area is by the superimposition of a grid (the coarseness of the grid mesh will depend on the level of accuracy required and the time available). Then: let the number of grid squares wholly within the cell be a, let the number of grid squares through which the cell boundary passes be b.

$$\text{the area of the cell} = \left(a + \frac{b}{2}\right) \times \text{the area of one grid square.}$$

Formula 2.6

The quantification of *shape* is much more difficult, since it has many facets which are not readily measured. Only one facet will be considered here—*compaction*.

A compact shape may be defined as one having, for a given area, a short boundary and a short distance between extremities. The most compact two-dimensional shape is a circle.

Perhaps the simplest measure of a cell's compaction is the length/breadth ratio (L/B):

$$L/B = \frac{\text{Length of long axis of cell}}{\text{Length of short axis of cell}},\qquad \textit{Formula 2.7}$$

where the long axis is the line joining the two points on the boundary which are furthest from each other in a straight line, and the short axis is the longest line that can be drawn perpendicular to the long axis between two other points on the boundary. In Figure 2.10, AB is the long axis, and CD the short axis. $L/B = 6/3 = 2$. The L/B for a circle will be 1, which is therefore its minimum value. The greater the value of L/B, the less compact the shape. On the other hand, a square will also have an L/B of 1, although it is considerably less compact than a circle. The length/breadth ratio is very easy to calculate, but is only a rough indicator of compaction.

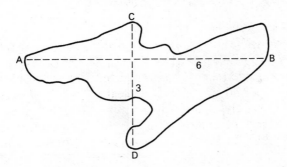

Figure 2.10. The length and breadth of an irregularly shaped area

A somewhat more refined measure of compaction is the

$$\textit{compaction index (C.I.):} = \frac{\text{area of cell being measured}}{\text{area of smallest inscribing circle}}$$

$$\textit{Formula 2.8}$$

A circular cell would have a C.I. of 1, which therefore indicates maximum compaction. The less compact the shape, the lower the value of C.I., which has a minimum value of 0. It should be noted that this index takes no account of the earth's sphericity, and should only be applied to areas where this can be ignored (up to, say, the size of Britain). An example will now be given to show how the C.I. is used.

Figure 2.11 shows the built-up area of Ruddington at different dates. (Ruddington is a large village in south Nottinghamshire, which had a flourishing domestic hosiery industry in the nineteenth century, and which is now in Nottingham's commuter belt. Its population growth between 1801 and 1971 is given in Table 3.2.)

The compaction index of Ruddington in 1770 is calculated as follows:

Area of built-up land in 1770 = $11 + \dfrac{24}{2}$ tenth-of-an-inch grid squares (using Formula 2.6).

Area of smallest inscribing circle = $\pi \times 4{\cdot}5 \times 4{\cdot}5$ similar grid squares (the radius being 4·5 tenth-of-an-inch units).

$$\text{C.I.} = \frac{11 + (24/2)}{\pi \times 4{\cdot}5 \times 4{\cdot}5} = \frac{23}{\pi \times 20{\cdot}25} = 0{\cdot}36.$$

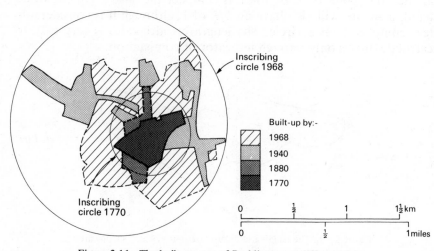

Figure 2.11. The built-up area of Ruddington at different dates

Similarly, the compaction index for 1968 is $211/(\pi \times 156) = 0{\cdot}43$. It will be noted that, although Ruddington has grown in the 200 years between 1770 and 1968, and has also changed in shape, the compactness of that shape was slightly greater at the end of the period than it was at the beginning. But the results will be seen to be more interesting if the following questions are considered and answered:

Consolidation

6. (a) Calculate the compaction index of Ruddington in 1880 and 1940. (b) Suggest why the compaction index for 1940 is so much lower than for 1880; and why it has recovered again in 1968. Is this a pattern you would expect to find in many built-up areas? If so, you could test your expectation (hypothesis) by carrying out a similar exercise on your own town or village.

2.5. THE MEASUREMENT OF DISCRETE AREAL DISTRIBUTIONS

Data are frequently published for administrative areas such as counties or parishes, which may then be mapped by choropleths. This section outlines two methods of summarizing such data (or maps). Both involve comparisons, and both, by comparing an actual set of areal data with a 'standard' distribution (which may also be 'actual' like the distribution of population, or may be hypothetical, like a perfectly even distribution), can be used as a measure of localization or spatial concentration.

2.5a. The Lorenz curve

We have seen how the Lorenz curve can be used to portray the numerical distribution of a variable among a number of categories, such as employment among industrial orders (p.25). It can equally be used to portray a spatial distribution, or to compare two distributions among spatial units. Consider the data in Table 2.8:

TABLE 2.8
Distribution of work force in selected industries among standard regions of U.K.
% of working population employed in standard regions 1966

Employment in:	N	YH	EM	EA	SE	SW	WA	WM	NW	Sc	N.I.	U.K.
A Agriculture	5·2	7·8	7·6	12·6	23·2	11·2	3·6	6·9	4·2	15·6	2·3	100
B Textiles	2·6	22·7	15·3	0·5	4·2	2·0	2·2	4·6	27·4	12·1	6·5	100
C Prof. and sc. services	5·2	7·9	4·9	2·7	36·0	6·6	4·6	8·3	11·6	10·0	2·2	100
D Total empl.	5·5	8·8	6·0	2·6	33·7	5·6	4·2	10·0	12·6	9·0	2·0	100
E Area	7·9	5·8	5·0	5·1	11·2	9·7	8·5	5·3	3·3	32·3	5·8	100

The figures in Table 2.8 represent spatial distributions; along any one row, each figure relates to a different area: location is one of the variables, so that these are 'space data', capable of being mapped, as some of them are in Figure 2.12.

Figure 2.12. Percentage distribution of agricultural work force among standard regions of U.K., 1966.

Lorenz curves have been drawn in Figure 2.13, to show the relationship between the data in each of the top three lines of the table (A, B, and C) and those in the fourth line (D).

In other words, each curve compares the distribution of employment in one industrial order with the distribution of total employment (which approximates to the distribution of the whole population). Notice the following points:

(a)   Both horizontal and vertical axes have interval scales along them, measuring cumulative percentages (unlike the Lorenz curve in Figure 1.9, in which only the vertical scale measures cumulative percentages).

(b)   The diagonal XY represents perfect correspondence between the distributions being plotted along the two axes, and the deviation of the curve from the diagonal is a measure of the dissimilarity of the two distributions. Thus, curve C, which is close to the diagonal, shows that the industrial order called 'Scientific and Professional services' is distributed similarly to the distribution of all employment,

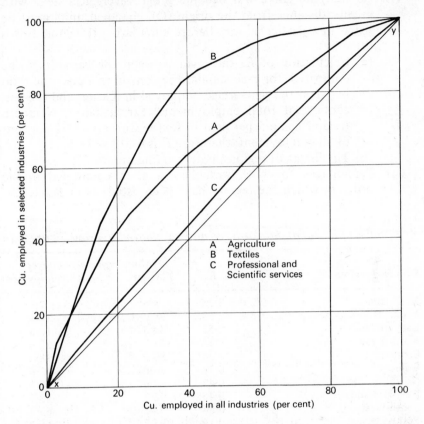

Figure 2.13. Lorenz curves to show regional distribution of selected industries in relation to regional distribution of all industries

and therefore to that of the whole population. This is a market-oriented industry—teachers, doctors, and lawyers need to be located near their 'markets'. By contrast, curve B represents a localized industry, relatively independent of the distribution of the whole population—suggesting that agglomeration economies or localized resources are more important to the textile industry than proximity to the domestic consumer.

Construction of a spatial Lorenz curve
Suppose we are drawing the Lorenz curve for textile employment against total employment (line B against line D in the table). The procedure is similar to that outlined in Chapter One for the non-spatial Lorenz curve, but there are two important differences:

(i) So that the curve will become progressively less steep with greater distance from the origin (O), the areal units must be listed in a particular way before cumulating the data. This is done by:

(a) Calculation of R, the ratio between the variables to be compared, for each standard region—in this case, the ratio between the percentage employed in textiles and the percentage of total employment.[4] For instance, Northern Ireland has 6·5 per cent of textile workers and 2 per cent of all employment; therefore R is $6·5/2 = 3·25$.

(b) The listing of regions in descending order of R.

(ii) Percentages for both variables are then cumulated in the order in which the regions have been listed, as in Table 2.9:

TABLE 2.9

Regional employment data ordered and cumulated to draw Figure 2.14

Standard region	R	Cumulative percentages Textiles	All employment
N. Ireland	6·5/2·0 = 3·25	6·5	2·0
Yorks. Humberside	22·7/8·8 = 2·58	29·2 (6·5 + 22·7)	10·8
East Midlands	2·55	44·5	16·8
North-West	2·17	71·7	29·4
Scotland	1·34	83·8	38·4
Wales	0·52	86·0	42·6
North	0·47	88·6	48·1
West Midlands	0·46	93·2	58·1
South-West	0·36	95·2	63·7
East Anglia	0·19	95·7	66·3
South-East	0·12	99·9	100·0

The figures in the last two columns are now plotted as in Figure 2.13, with the 'standard' data (i.e. the data against which the distribution is compared to measure localization—in this case, the data for all employment) along the horizontal axis.

Consolidation

7. (a) Construct Lorenz curves for.

(i) agriculture and area (lines A and E in Table 2.8)

(ii) all employment and area (lines D and E).

(b) What does each of these Lorenz curves show? What conclusion can be drawn by comparing them?

Index of concentration
An index of concentration, comparable to that given in connection
with the non-spatial Lorenz curve (Formula 1.6) can be calculated,
using a similar but not identical procedure. This procedure is best
followed in conjunction with Figure 2.14.

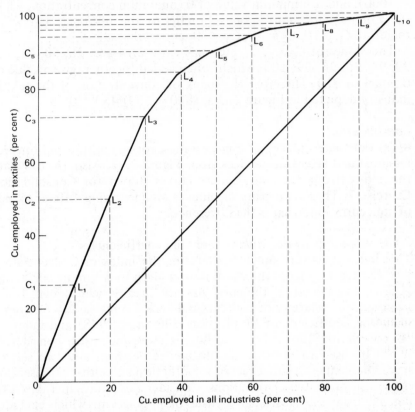

Figure 2.14. Lorenz curve for U.K. textile industry, to show how index of concentration
is derived

The steps are:
  (i) mark off ten equally spaced points along the horizontal axis;
  (ii) project vertically from these points to the Lorenz curve at
       L_1, L_2, etc.;
  (iii) project from L_1, L_2, etc. to the vertical axis at C_1, C_2, etc.,
        and note the values of C_1, C_2, etc. on the vertical scale;
  (iv) add the ten values C_1, C_2, etc. to give C. (This is equivalent
       to A in Formula 1.6.) The index of concentration is now
       given by

$$I = \frac{C-550}{1000-550} = \frac{C-550}{450}. \qquad\qquad \textit{Formula 2.9}$$

This will give a maximum value of 1 (maximum concentration, when the Lorenz curve deviates as much as possible from the diagonal, and when therefore $C_1 = C_2 = C_3 = C_4$ etc. $= 100$, so that $C = 10 \times 100 = 1000$); and a minimum value of 0 (minimum concentration, when the Lorenz curve corresponds exactly to the diagonal, and when therefore $C_1 = 10$, $C_2 = 20$, $C_3 = 30$, etc. so that $C = 550$).

The values of C_1, C_2, C_3, etc. on Figure 2.14 are: 26, 50, 72, 85, 90, 94, 96, 98, 99, 100, which, when added together, give a value of C equal to 810. Therefore the index of concentration of the textile industry depicted in Figure 2.14 is $(810 - 550)/450 = 0.58$.

Consolidation
8. (*a*) Calculate indexes of concentration for agriculture and for professional and scientific services from Figure 2.13. Also, (*b*) calculate the index from the two Lorenz curves drawn for Consolidation Exercise 7. These two pairs of indexes are measuring different kinds of concentration; what is the difference?

2.5b. The index of dissimilarity (or Gini coefficient)
This index does very much the same as the Index of Concentration just discussed. It lacks the visual impact given by a Lorenz curve, but is easier to calculate if the curve has not already been drawn. It has been used in a variety of contexts and been given a variety of names, including coefficient of localization (Florence, 1948) and index of change (Smith, 1969). Like the Lorenz curve index just considered, it can be used to measure the difference between any two sets of paired percentages, either real or hypothetical and either representing numerical or spatial distributions. Consider the figures in Table 2.10 derived from the Australian Census of Population (which excludes full-blooded Aboriginals from its count):

TABLE 2.10
% of population in Australian states at different dates

Pop.	N. Terr.	N.S.W.	Victoria	South A.	West A.	Queens.	Tasmania
1881	0·15	33·32	38·29	12·30	1·32	9·49	5·14
1921	0·07	38·69	28·17	9·11	6·12	13·91	3·93
1961	0·26	37·84	27·88	9·22	7·01	14·45	3·33
Area	17·62	10·42	2·96	12·79	32·85	22·45	0·88

The figures in the first three rows show how the *relative* importance of each state has altered in terms of its population—though no indication is given of actual numbers. By calculating the index of dissimilarity for the data of 1881 and 1921, and then for 1921 and 1961, we can compare the changes that occurred in the two periods of forty years. The formula for the index of dissimilarity is:

either $\Sigma(p - q)$ where p is greater than q,
or $\Sigma(q - p)$ where q is greater than p, *Formula 2.10*

and where p and q are the sets of percentages to be compared. If both sets of percentages add up to exactly 100, then both formulae will give exactly the same result. The index will range from 100 (maximum dissimilarity) to 0 (complete similarity).

If we apply the formula to the data for 1881 and 1921 (which we can call p and q respectively), then the states in which p is greater than q are Northern Territory, Victoria, South Australia, and Tasmania. Therefore, $\Sigma(p - q)$ for these states is:

$$(0.15 - 0.07) + (38.29 - 28.17) + (12.30 - 9.11)$$
$$+ (5.14 - 3.93) = 14.60.$$

Applying the formula to a comparison of 1921 and 1961, the answer is 1·54. In other words, a considerable change in the relative importance of the states occurred during the earlier period, and practically none during the second. Can you suggest why?

Having used the index of dissimilarity as a measure of change, let us now use it as a measure of spatial concentration. If population were distributed perfectly evenly, then the percentage in each state would equal the percentage of the total national area occupied by each state. Therefore, by comparing population distribution with the distribution of land between the states, we have a measure of population concentration: the greater the dissimilarity, the greater the concentration. This is done for 1961: p is line three in Table 2.10, and q is line four. $\Sigma(p - q)$ where p is greater than q is:
$(37.84 - 10.42) + (27.88 - 2.96) + (3.33 - 0.88) = 54.79$.

Consolidation
9. Calculate similar measures of spatial concentration for 1881 and 1921. Has the population of Australia become more or less concentrated during the period 1881 to 1961? Can you explain your answer?

2.5c. Limitations of the Lorenz curve and index of dissimilarity
The measures described in the foregoing sections make use of data
that have been aggregated for areas such as British Standard Regions
and Australian states. Any variability occurring within these areas is
therefore ignored, reducing the amount of spatial variation (or concentration) that can be measured. In general, the greater the number
of subdivisions for which data are given, the greater the measure of
variation (or concentration).

This is strikingly illustrated in Figure 2.15, where indexes of
dissimilarity (here called indexes of population concentration) for
the distribution of United States population have been graphed,
using five systems of areal subdivision. It will be seen that systems
with a large number of small subdivisions (such as counties) yield
higher indexes than those with fewer but larger subdivisions (such as
states).

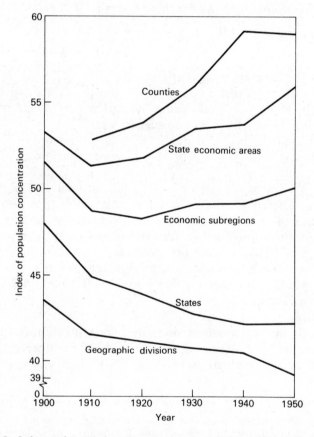

Figure 2.15. Indexes of population concentration for various systems of areal subdivision
of the United States, 1900 to 1950 from Isard (1960) © 1960 M.I.T. Press

Perhaps more interesting is the fact that the graphs show different trends. When the country is divided into states, there is a downward trend in the index, implying that the population has become more evenly spread between states (no doubt due to the general westward shift of people into the sparsely inhabited states beyond the Mississippi). But when the country is divided into counties, there is an upward trend in the index, showing that people became less evenly distributed between counties (reflecting the movement from rural to urban counties within the states).

This example serves to remind us that, (i) indexes such as those outlined in sections 2.5a and 2.5b are only comparable when related to the same areal subdivisions—they should not, for instance, be used to compare the distribution of population among the states of Australia and those of the U.S.A.; and (ii) a distribution pattern may be too complex for even a single facet, such as spatial concentration, to be adequately summarized by a single index—which takes us back to the concept mentioned at the beginning of this chapter, that of the statistical profile. No single figure can do justice to a spatial distribution.

2.6. THE MEASUREMENT OF SPATIALLY CONTINUOUS DISTRIBUTIONS
Where a spatial distribution is mapped by isopleths, a continuum is implied, analogous to a topographic surface in which every point has an altitude and gradient. It has long been customary to describe isobar patterns with topographic terms such as ridge, trough, and col. The practice has now been extended to all isopleth patterns, and one reads of 'surfaces' of cost, land value, accessibility, and population density—though not all of these depict spatially continuous phenomena. Whether phenomena are continuous or not, however, it has been

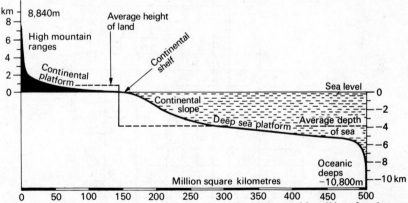

Figure 2.16. Hypsographic curve, showing the areas of the earth's solid surface between successive levels from Holmes (1940)

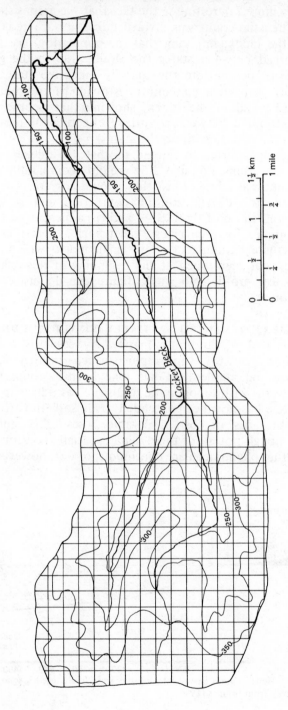

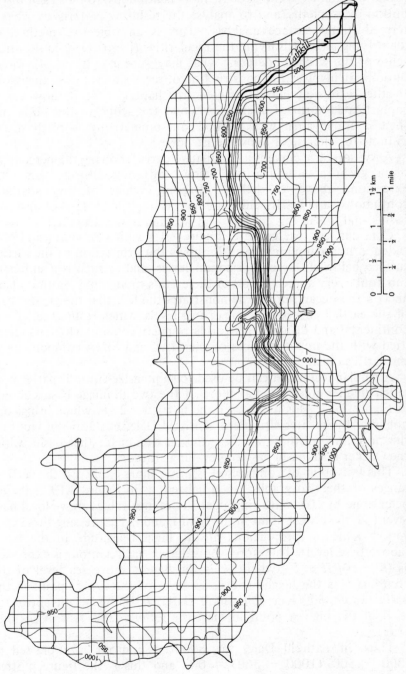

Figure 2.17. (a) and (b) Contour patterns of the Cocker Beck and Lathkill drainage basins

found profitable to think of them as continuous, to plot their distribution by isopleths, and to analyse the resulting patterns by what is termed 'trend surface analysis'. This is an objective method of identifying and quantifying the regularities (i.e. patterns in the strict sense of the word) underlying what may appear to be chaotic distributions. It involves the construction of regression surfaces analogous to the regression lines described in Chapter 8. It is, however, a sophisticated form of analysis, beyond the scope of this book. We shall consider a simpler method of summarizing isopleth distributions—by the use of hypsometric curves.

A hypsometric curve is a cumulative curve, showing the percentage of a given area above (or below) differing altitudes—and, by extension, above or below differing values of any spatially continuous variable. Its chief use has been in the analysis of drainage basins, and it is in this context that we shall now discuss it.

Consider first the graph in Figure 2.16, which A. Holmes (1944) called a 'hypsographic curve'. It shows the proportion of the earth's surface below different horizontal levels, and as such is a summary (no doubt very approximate) of the earth's topography, both on land and on the ocean floor. It demonstrates the fact that the greater part of the earth's surface is at one of two levels, which Holmes called the continental and deep-sea platforms respectively, and which he identified with the upper surfaces of the SIAL and SIMA components of the earth's crust.

A similar technique can be used to summarize contour patterns of much smaller areas, such as those of the two drainage basins shown in Figure 2.17. This has been done in Figure 2.18, which brings out not only the greater altitude of Lathkill Dale, but also the fact that the prevailing altitudes are near the maximum for that basin, while the Cocker Beck basin has no prevailing altitudes.

This last point of comparison is made more explicit if the vertical ranges of the two graphs are equalized, as in Figure 2.19. This has been done by: (i) converting all altitudes into values above local base level (i.e. the lowest point in the basin); and (ii) expressing these converted values as proportions of the highest altitude in the basin (above base level). Therefore, the value of each contour is expressed as $(h - b)/(H - b)$, where: h is the height above sea level of the contour; b is the height above sea level of the lowest point in the basin (about 500 feet in Lathkill Dale); H is the height above sea level of the highest point in the basin (about 1000 feet in Lathkill Dale).

Thus, in Lathkill Dale, the 800-foot contour is expressed as $(800 - 500)/(1000 - 500) = 0.6$, and this is the figure plotted

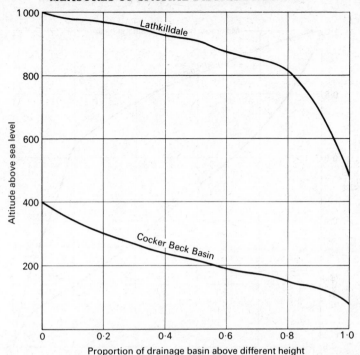

Figure 2.18. Hypsographic curves of the Cocker Beck and Lathkill drainage basins

against a vertical scale ranging from 0 to 1. The horizontal scale measures the area of the basin above a given countour (*a*), as a proportion of the total area of the basin (*A*)–i.e. *a*/*A*, which will also range from 0 to 1.

The graphs so drawn are called Hypsometric curves. As they both start and finish at the same points, comparisons are easier to make. The over-all 'convexity' of Lathkill Dale as compared with the Cocker Beck basin is striking. This does not mean that most of the slopes of Lathkill Dale are convex, but that reduction of the basin's landscape to its present base level is not so far advanced there as in the Cocker Beck basin.

The comparison can be taken a step further by measuring the area under each Hypsometric Curve. This is called the hypsometric integral, and is usually expressed as a proportion of the area of the rectangle of which the two axes are adjacent sides. If, in Figure 2.19, the length of each axis is 10 units, then the area of the rectangle is 100 square units, the area under the Lathkill Dale curve is about 70, and under the Cocker Beck curve about 50 square units; therefore

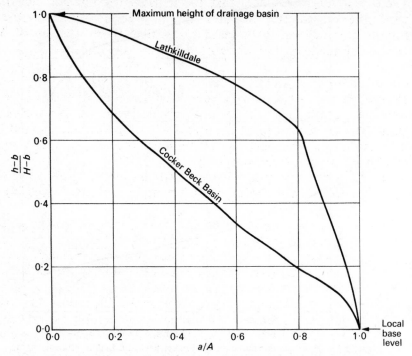

Figure 2.19. Hypsometric curves of the Cocker Beck and Lathkill drainage basins

the hypsometric integral for Lathkill Dale is 0·7 (or 70 per cent), and for the Cocker Beck basin is 0·5 (or 50 per cent).

The hypsometric curve and integral are useful descriptive summaries and criteria for the comparison of drainage basin relief. They may also be useful tools in the analysis of drainage basin development. For instance, it has been observed that

in actively eroding small badland basins, after about 25 per cent of the theoretical mass of the basin has been removed (i.e. hypsometric integral of less than 75 per cent), the basin relief ratios and mean stream gradients remain essentially constant until at least 80 per cent has been removed (hypsometric integral less than 20 per cent). Similarly, it appears that once a high point of a drainage basin is involved in the general degradation, at a hypsometric integral of 60 per cent or less, the integral more-or-less stabilizes itself, irrespective of the absolute relief, as the basin is geometrically transformed. . . .

Thus, within a comparatively small range of hypsometric integral lies a wide range of basin relief, implying that, when applied to a single basin whose relief is being decreased through time, there is a long equilibrium (or mature) stage characterized by a more-or-less stable ratio of the hypsometric integral. (Chorley and Kennedy (1971), pp.229–30.)

There is a suggestion here that the hypsometric integral, together with other quantitative measures, may provide more precise criteria for the classification of drainage basins than those used by W. M. Davis in his identification of youth, maturity, and old age.

Consolidation
10. Select a small stream in your local area (about 2 miles from source to confluence with a larger river). From the $2\frac{1}{2}$-inch map, mark out the watershed of the area drained by your stream and its tributaries, and trace off the contours (if the area is one of high relief, trace alternate contours only). Then:

   (i) Note the lowest and highest altitudes of the basin (H and b), and express the value of each contour you have traced (h) as $(h-b)/(H-b)$.

   (ii) Superimpose a grid to find the area (a) enclosed by the watershed and each contour in turn. (To avoid repeating measurements, start with the highest contour and add on areas as you work down.) Express each value of a as a proportion of the total area of the basin (A)–i.e. a/A.

   (iii) Plot a hypsometric curve as in Figure 2.19, and calculate the hypsometric integral. How does your result compare with those of Lathkill Dale and the Cocker Beck basin? Does your basin appear to be at the equilibrium (or mature) stage of development suggested by Chorley and Kennedy?

3
TIME SERIES

3.1. INTRODUCTION

ALTHOUGH geography is primarily concerned with spatial variations, and history with change through time, geographers cannot ignore the time dimension (nor historians that of space). The geographer's interest in time stems from two considerations:

(i) Like other scientists, he is concerned with explaining what he observes; explanation involves the study of process and sequence, and these occur through time. The explanation of the present is rooted in the past. 'That which was is now; and that which is to be hath already been' (Ecclesiastes 3:15).

(ii) Change through time is an essential characteristic of place, and different rates of change in different places constitute a major spatial variable.

Time series have been defined as 'successive observations of the same phenomenon over a period of time' (Alder and Roessler, 1964). Much time series analysis involves the use of graphs, in which the horizontal scale measures time, and the vertical scale the magnitude of the variable being studied. We therefore begin this chapter with a brief consideration of graphs.

3.2. GRAPHS

Time series graphs are of two kinds: line graphs and histograms. Strictly speaking, line graphs should only be used where each point on the graph relates the magnitude of the variable under consideration to a *point* in time—e.g. graphs of temperatures or population growth. Where magnitudes are related to *periods* of time—e.g. monthly rainfall or annual birthrate—histograms are more appropriate. Nevertheless, line graphs are often used for time-period data, as they are quick to draw, and several lines can be superimposed to facilitate comparison of trends. It should be recognized, however, that such trend lines for *period*-time data cannot be used to interpolate magnitudes for *points* in time, or for different periods of time. One cannot read momentary, daily, or monthly birthrates off a graph depicting annual birthrates.

Furthermore, where data relating to points in time are plotted, common sense is needed in constructing and interpreting the line joining the points. Where there are reasonable grounds for believing that magnitudes have changed steadily between observations (as with national populations) a smooth curve may be appropriate and may be used to interpolate intermediate values. But where considerable fluctuations are known or suspected between observations (as is the case with the livestock data discussed in the next section), line graphs cannot be used to interpolate intermediate values. They serve only to depict the trend or rate of change between observations, and points should be joined by straight lines rather than by curves.

We shall now consider three aspects of time series analysis: (i) measurement of growth and decline; (ii) identification of trends and fluctuations; (iii) projections and predictions.

3.3. GROWTH AND DECLINE
This discussion hinges on the distinction between numerical and proportional change. Numerical change is, in part, a function of the units of measurement used, and misleading comparisons can be made if this fact is not borne in mind. Consider the graphs in Figure 3.1.

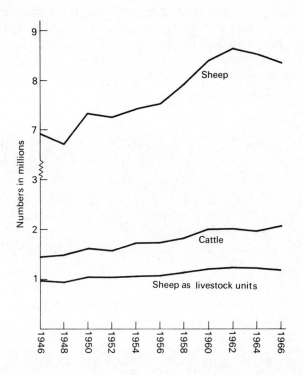

Figure 3.1. Numbers of sheep and cattle in Scotland, 1946 to 1966

If we compare the number of sheep (top graph) and number of cattle, it appears that (i) sheep numbers have, on the whole, grown faster than those of cattle, and (ii) sheep numbers have been more variable from year to year. But it is hardly realistic to compare numbers of sheep and cattle directly. If a numerical comparison is required, a common unit of measurement is necessary, such as standard labour requirements, expressed in man-days (Chapter One, p.36), or livestock units (based on the average food consumption and value of different kinds of animal). When the figures are expressed in Livestock Units, the graph for cattle remains the same (since one Livestock Unit = one head of cattle), but the sheep graph is lower, flatter, and contains less fluctuation than that for cattle (one livestock unit = seven sheep).

It is generally more meaningful to compare proportional rather than numerical change, for then units of measurement do not matter. For instance, the proportional growth of the sheep population of Scotland shown in Figure 3.1 was the same, whether calculated from numbers of sheep or from numbers of livestock units. Between 1956 and 1960 it was 11·7 per cent, a figure which is directly comparable with the proportional growth in the number of cattle over the same period—14·8 per cent. But the gradients of the graphs in Figure 3.1 reflect numerical change and not proportional change, and comparisons of proportional change cannot be made by consulting these graphs.

3.4. INDEX NUMBERS

One method of measuring and depicting proportional change is by the use of index numbers. A 'base year' is chosen, and all figures for other years are expressed as percentages of the figures for the base year. The latter are therefore equal to 100.

The raw data from which Figure 3.1 is derived are set out in Table 3.1, alongside index numbers with 1946 as base year. The index numbers are graphed in Figure 3.2.

An example to show how index numbers are calculated follows: no. of cattle in 1950: 1 616 000; no. of cattle in 1946: 1 472 000. Therefore, the 1950 figure expressed as an index number with 1946 as base year is given by:

$$\frac{1\,616\,000}{1\,472\,000} \times 100 = 110$$

The advantage of using index numbers is that they are independent of the initial magnitude of the data, and of the units in which they are measured. They express each figure as a percentage of the

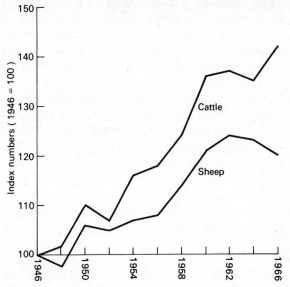

Figure 3.2. Index numbers of sheep and cattle in Scotland, 1946 to 1966 (1946 = 100)

TABLE 3.1
Numbers of sheep and cattle in Scotland, 1946—66, with index numbers (base year = 1946)

| | Cattle | | Sheep | |
	Numbers (X 1000)	Index nos.	Numbers (X 1000)	Index nos.
1946	1472	100	6954	100
1948	1499	102	6731	98
1950	1616	110	7337	106
1952	1576	107	7273	105
1954	1710	116	7429	107
1956	1736	118	7525	108
1958	1820	124	7929	114
1960	2003	136	8407	121
1962	2017	137	8639	124
1964	1990	135	8531	123
1966	2091	142	8377	120

base year, so that comparisons of growth and decline can be made easily. Figure 3.2 shows clearly that cattle have increased pro-portionally more rapidly than sheep, although the reverse was true in terms of numerical growth. It is also easy to read off proportional change: an index number of 136 means a growth of 36 per cent since base year; an index number of 217 means a growth of 117 per cent; one of 88 means a decline of 12 per cent. Finally, the base year does not have to be the first year of the time series. If another year is chosen, then graphs for different variables (such as sheep and cattle) will meet (and perhaps cross) at base year, because all index numbers for that year equal 100.

One limitation of index numbers is that they do not show proportional change between any two years in a time series, but only between one year and the base year. For instance, the index numbers for sheep and cattle in Scotland both went up by six points between 1956 and 1958 (from 108 to 114, and from 118 to 124 respectively). The slopes of the two graphs in Figure 3.2 are the same between these dates. But six points on 108 represents a rise of nearly 6 per cent in sheep numbers, while six points on 118 represents only a 5 per cent rise in cattle numbers.

Consolidation
1. Using Index numbers, compare the twentieth-century growth of Nottingham and Ruddington populations from the figures in Table 3.2. Make 1931 base year.

3.5. LOGARITHMIC SCALES
If one wishes to show proportional change between any and all points in a time series by graph, a logarithmic scale must be used. A logarithmic scale differs from an ordinary arithmetic scale in the way shown in Figure 3.3.

On an arithmetic scale, a given *numerical* difference is shown by a constant interval; on a logarithmic scale, a given *proportional* difference is shown by a constant interval. If a time series is to be depicted logarithmically, it is usual to use 'semi-logarithmic' graph paper, in

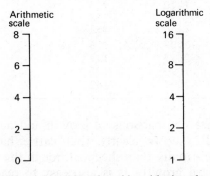

Figure 3.3. Arithmetic and logarithmic scales

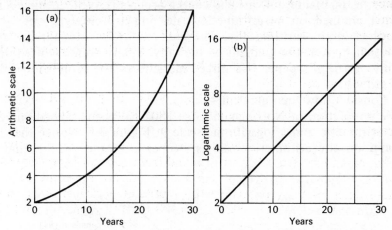

Figure 3.4. Two graphs depicting a population that is doubling every ten years
(a) On an arithmetic scale, (b) On a logarithmic scale

which the vertical scale is logarithmic, and the horizontal scale, measuring time, is arithmetic. (Graph paper with both scales logarithmic is called 'double-logarithmic'.) Figures that are increasing at a constant proportional rate per unit of time (e.g. doubling every ten

TABLE 3.2
Population statistics for Nottingham and Ruddington 1801 to 1961

Census year	Ruddington	Nottingham
1801	868	28 801
1811	1017	34 030
1821	1138	40 190
1831	1428	50 220
1841	1835	52 164
1851	2181	57 407
1861	2283	74 693
1871	2436	86 621
1881	2638	186 575
1891	2370	213 877
1901	2493	239 743
1911	2771	259 901
1921	2877	262 624
1931	3064	276 189
1941	No Census during War	
1951	4530	306 008
1961	5185	311 899
1971	6838	300 630

years, as in Figure 3.4) are shown as a progressively steepening curve when graphed on an arithmetic scale, and as a straight line when graphed semi-logarithmically. The gradient of the straight line (or, for that matter, the gradient of the curve at any point) plotted onto semi-logarithmic paper is a direct measure of the proportional rate of growth.

Consider now the figures in Table 3.2.

The population of Nottingham since 1801 has been graphed on both an arithmetic and a logarithmic scale in Figure 3.5. Notice that the graph on the arithmetic scale is steeper after the 1877 boundary

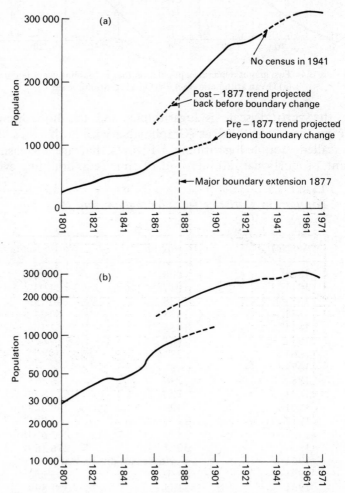

Figure 3.5. (a) Population growth in Nottingham, 1801 to 1971, on arithmetic scale
            (b) Population growth in Nottingham, 1801 to 1971, on logarithmic scale

change than it was before, indicating that the numerical increase following the boundary change was greater than before. But there is nothing very remarkable in a large population (post-1877) experiencing a greater numerical increase than a smaller population pre-1877), as we saw in the example of the Scottish sheep and cattle. What is much more significant is the proportional rate of growth, and the semi-logarithmic graph's pre- and post-1877 gradients show that this had not altered substantially.

Now look at Figure 3.6, which shows two graphs, representing data of very different orders of magnitude—a large city and a village

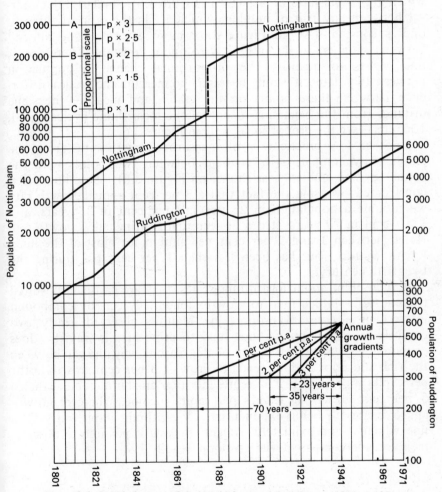

Figure 3.6. Population growth in Nottingham and Ruddington, 1801 to 1971

population. Because the logarithmic scale shows proportional and not numerical differences by a constant interval, the gradients of the two graphs are directly comparable. It is clear that Nottingham and Ruddington grew at about the same rate from 1801 to 1831 (the graphs being roughly parallel here); between 1831 and 1851, Ruddington grew faster (though it increased by less than a thousand, while Nottingham increased by over seven thousand); from 1851 to 1881 Nottingham grew faster; and so on.

Notice that the interval on the vertical scale between 1000 and 2000 is the same as that between 3000 and 6000, 4000 and 8000, 40 000 and 80 000—in fact, it is the same for any pair of numbers which are in the ratio of 2:1 (BC in Figure 3.6). Similarly, the interval between any pair of numbers which are in the ratio of three to one is constant (AC in Figure 3.6); more generally, the interval between numbers in the ratio of n to one is constant. Therefore, if one wishes to quantify the proportional change in population between two dates, it is only necessary to measure the vertical difference between these points on the graph, and measure the same distance from the beginning of one of the logarithmic cycles (1, 10, 100, etc.)—see the proportional scale in Figure 3.6.

Another less precise way of measuring growth rates from the graph is by comparing its gradient between two points with the annual growth gradients (as shown on Figure 3.6). These have been constructed by using the fact that a 3 per cent annual growth rate will double a population in 23 years; 2 per cent in 35 years, and 1 per cent in 70 years. It can be seen that Nottingham's growth rate varied between 1 and 2 per cent per annum for most of the nineteenth century, but has been less than 1 per cent p.a. during the twentieth century.

A final point about logarithmic scales: they are particularly suited to time series which tend to grow exponentially (i.e. by compound interest or a constant proportion). Malthus observed nearly two centuries ago that populations tend to grow in this way. So does economic growth, which is one of the reasons why the gap between the rich and poor nations tends to grow—a 3 per cent growth in the per capita income of Britain represents a far greater increment than a 3 per cent growth in India's per capita income. Wherever growth approximates to being exponential, a logarithmic graph will yield a line which is approximately straight—making it possible to observe, measure, and compare deviations more readily.[1]

Consolidation
2. Draw graphs of the data in Table 3.1 on semi-logarithmic graph paper.

3.6. TRENDS AND FLUCTUATIONS

A variable's change through time may be one of steady growth or decline; or it may be characterized by wild and apparently random fluctuations from one observation to the next. In fact, three components are often recognized in a time series:

   (i) the over-all or long-term trend—sometimes referred to as the secular trend;

  (ii) periodic fluctuations of a rhythmic nature—associated with daily, seasonal or other cyclic variations;

 (iii) irregular or random variations.

Consider the data in the first two columns of Table 3.3, which show the pig population of Britain from 1916 to 1939. These have been graphed in Figure 3.7. It is clear that there has been an over-all increase in the pig population during this period (what may be called an upward secular trend), but there have been wide fluctuations from year to year.

TABLE 3.3
The Pig Population of Great Britain, 1916 to 1939

Year	No. of pigs × 1000	Semi-averages	Five-year moving averages
1916	2168		
1917	1919		
1918	1697		1915
1919	1798		1983
1920	1994		2059
1921	2505		2242
1922	2299	– – – – – – 2313 – – – – – – – – –	2528
1923	2612		2658
1924	3228		2597
1925	2644		2675
1926	2200		2747
1927	2692		2575
1928	2971	– – – mid-point of time series – – – – –	2508
1929	2367		2625
1930	2310		2723
1931	2783		2743
1932	3185		2933
1933	3069		3234
1934	3320	– – – – – – 3195 – – – – – – – – –	3438
1935	3813		3528
1936	3804		3627
1937	3635		3666
1938	3564		
1939	3515		

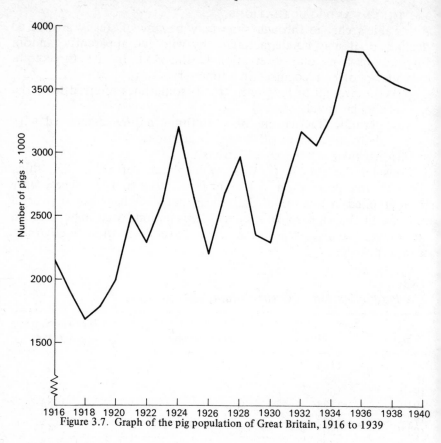

Figure 3.7. Graph of the pig population of Great Britain, 1916 to 1939

Let us be more precise. As a first step, a trend line can be drawn. The easiest method is by the use of semi-averages. The average value of each half of the time series is calculated, and assigned to the middle year of that half. (Column 3, Table 3.3.) The line drawn through these points is the trend line. The original graph has been redrawn, together with the trend line, in Figure 3.8. It is now easier to identify the fluctuations about the secular trend. They can be measured, and plotted (as in Figure 3.9).

The question then arises: are these fluctuations about the trend line entirely random, or do they follow a pattern? If the positive and negative deviations from the trend line are noted, the result is as follows: $+----+-+++--+----+-+++++-$. Clearly the sequence is not perfectly regular, but nor does it appear to be entirely irregular. Perhaps it is compounded of both regular (rhythmic) and irregular (random) elements.

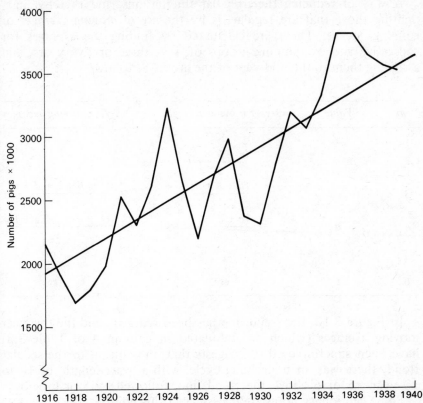

Figure 3.8. Graph and trend line to show the changing pig population of Britain, 1916 to 1939

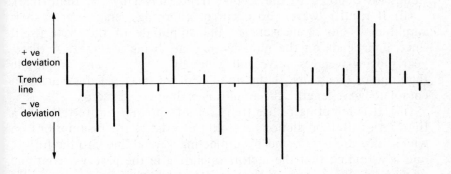

Figure 3.9. The pig population of Britain, 1916 to 1939: Deviations of annual figures from the secular trend

A way of reducing the irregular fluctuations, and thereby high-lighting those that are regular, is by the use of *moving averages* or running means. These are calculated by finding the averages for successive and overlapping groups of, say, three or five years, and assigning them to the mid-year of the group, as follows:

Year	Value	3-year moving averages	5-year moving averages
1	a		
2	b	$\dfrac{(a + b + c)}{3}$	
3	c	$\dfrac{(b + c + d)}{3}$	$\dfrac{(a + b + c + d + e)}{5}$
4	d	$\dfrac{(c + d + e)}{3}$	$\dfrac{(b + c + d + e + f)}{5}$
5	e		
6	f	etc.	etc.

In Figure 3.10, the trend line has been redrawn, and the five-year moving averages (which are tabulated in column 4 of Table 3.3) have been superimposed. It suggests that, in addition to the secular trend, there was an underlying cycle, with a 'wave-length' of 10 to 12 years (5 or 6 above the trend line, followed by 5 or 6 below).

Some cautionary notes are necessary at this point:

(i) Not all time series contain a cyclic element. But moving averages may, still, be useful in smoothing out erratic variations between successive observations, and thereby revealing short-term trends (as opposed to the secular trend shown by the trend line).

(ii) If regular 'waves' do occur in a moving-averages curve, their amplitude is not a measure of the amplitude of the cycle itself, since it depends on the number of years being averaged. A three-year moving-averages curve will generally have a greater amplitude than one of five years. If the number of years averaged is equal to a complete 'wave-length', the amplitude will approach zero.

(iii) It is tempting to use trend lines to predict future values of a time series. But predictions are only as valid as the assumptions on which they are based, and by projecting a trend line into the future, one is assuming that the factors operating in the past will continue unchanged. By projecting the trend line for the pig population of Britain to 1944, one reaches a figure of 4 million; the actual figure for that year was 1·46 million. The War had completely altered the circumstances affecting pig breeding in Britain.

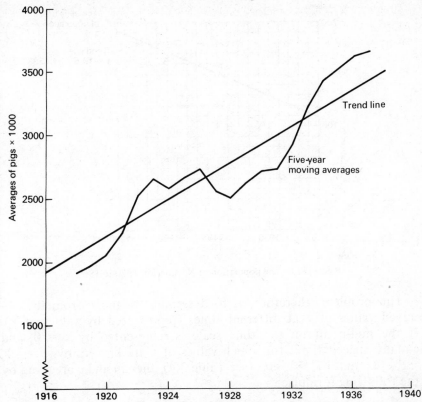

Figure 3.10. The pig population of Britain, 1916 to 1939: Trend line and five-year moving averages

3.7. TREND LINES BY LEAST-SQUARES METHOD

A trend line should be drawn in such a way that deviations of individual values from it are minimized. The semi-averages method already described (p.96) is quick, and, for many purposes, adequate. A longer, but more rigorous method is one called the 'least-squares' method, because it minimizes the sum of the squares of the differences between observed values and corresponding trend-line values. It involves first finding the equation of the 'least-squares' line, and then plotting it.

All straight-line graphs have equations of the form $y = mx + c$, where x and y are variables, and m and c are constants which determine the gradient of the line, and where it cuts the y-axis respectively. Every line has a unique combination of values of m and c, and when these are known, the line can be drawn by plotting two points and ruling through them.

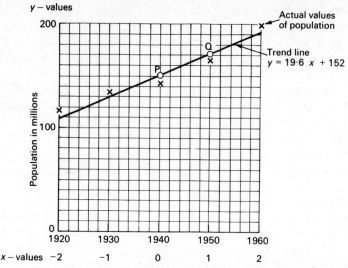

Figure 3.11. The population of N. America, 1920 to 1960

The problem therefore, is to determine m and c from the observed values of y at different times (represented by values of x). If the mid-point on the time scale is represented by $x = 0$, and equally spaced times for which values of y are known, by $x = -1$, -2, etc., and $+1$, $+2$, etc. (see Table 3.4), then m and c are given by the following formulae:

$$m = \frac{\Sigma(xy)}{\Sigma(x)^2} \quad \textit{Formula 3.1} \qquad c = \frac{\Sigma y}{n} = \bar{y} \quad \textit{Formula 3.2}$$

An example, using population figures for North America between 1920 and 1960, will demonstrate how the calculations are carried out. All the figures are set out in Table 3.4:

TABLE 3.4

The population of N. America, 1920–1960

Year	No. of decades from mid-year (x)	Population in millions (y)	x^2	xy
1920	-2	117	4	-234
1930	-1	134	1	-134
1940	0	144	0	0
1950	1	166	1	166
1960	2	199	4	398
		$\Sigma y = 760$	$\Sigma x^2 = 10$	$\Sigma(xy) = 196$

$$m = \frac{196}{10} = 19 \cdot 6 \qquad\qquad c = \frac{760}{5} = 152$$

Therefore, the equation of the trend line is $y = 19\cdot6x + 152$. (Note: the conversion of time to x-values in column 2 is to ease calculation; if there is an even number of observations, the mid-year is taken as half-way between the two middle years, and the x-values are $0\cdot5$, $1\cdot5$, $2\cdot5$, etc.)

The drawing of the line simply requires the plotting of two points which satisfy the equation that has been found, and the ruling of a straight line through them. The easiest points to find (P and Q on Figure 3.11) are:

   (i)   when $x = 0$, $y = m \cdot 0 + c = c = 152$
   (ii) when $x = 1$, $y = m \cdot 1 + c = m + c = 171\cdot6$.

Consolidation
3. Draw a semi-averages trend line and a three-year moving-averages curve for *either* the first *or* the second half of the time series in Table 3.3. How do they compare with Figure 3.10?
4. Calculate the equation and draw the graph of the least-squares trend line for the data used in Consolidation Exercise 3. How does this compare with the semi-averages trend line?

3.8. TREND LINES FOR EXPONENTIAL (i.e. LOGARITHMIC) SERIES
The discussion of trend lines has so far assumed that trends are approximately linear (2, 4, 6, 8). Many time series tend to grow exponentially (2, 4, 8, 16) as we have seen (p.94), and for these linear trend lines are inappropriate. Consider the hypothetical data in Table 3.5, which have been plotted in Figure 3.12. Trend lines have been calculated and plotted by both the semi-averages and least-squares methods. Neither is very satisfactory, because the data conform more closely to an exponential than to a linear series. A curved trend line would be more appropriate, but it is difficult to draw; and to calculate an equation for a least-squares curve would be even more difficult.

TABLE 3.5
A hypothetical time series

Year	Number
1	1
2	3
3	10
4	20
5	80

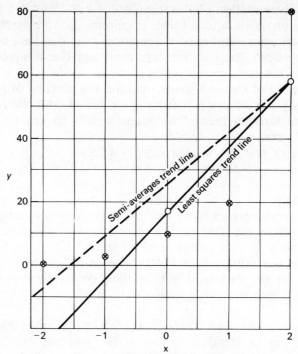

Figure 3.12. Linear trend lines on a non-linear time series

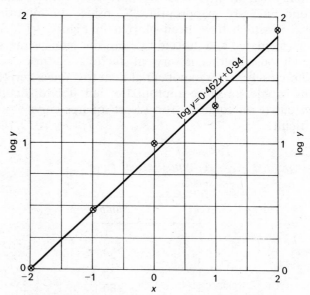

Figure 3.13. Linear trend line on non-linear time series after logarithmic transformation

A solution to this problem is to transform the data so that they do approximate more closely to a linear series. It has already been shown (p.94) that if a variable is growing exponentially, then its logarithm grows in a linear fashion. Therefore, by plotting the logarithms of the numbers in Table 3.5, we have a time series which is approximately linear, and for which a straight trend line is more meaningful (Figure 3.13). The trend line can be determined in exactly the same ways as already described (using either semi-averages or least-square method), simply substituting the logs of the numbers (y-values) for the numbers themselves. The working for the least-squares equation is set out below (Table 3.6). The equation is now of the form $\log y = mx + c$, where $m = \Sigma(x \log y) / \Sigma x^2$ and $c = (\Sigma \log y)/n = \overline{\log y}$—i.e. wherever y occurred previously $\log y$ now takes its place.

TABLE 3.6
Logarithmic transformation of hypothetical time series

Year	No. of years from mid-year x	Number y	x^2	$\log y$	$x \log y$
1	−2	1	4	0·0	0·
2	−1	3	1	0·48	−0·48
3	0	10	0	1·0	0·
4	1	20	1	1·3	1·3
5	2	80	4	1·9	3·8
			$\Sigma x^2 = 10$	$\Sigma \log y = 4\!\cdot\!68$	$\Sigma(x \log y) = 4\!\cdot\!62$

Therefore, $m = 4\!\cdot\!62/10 = 0\!\cdot\!462$; $c = 4\!\cdot\!68/5 = 0\!\cdot\!94$. The required equations is therefore: $\log y = 0\!\cdot\!46x + 0\!\cdot\!94$. This has been plotted, together with the actual values of $\log y$, in Figure 3.13. The line clearly fits the points better than either line in Figure 3.12.

Consolidation
5. The following table (3.7) sets out the population of England and Wales for each census year from 1841 to 1901, together with some of the calculations necessary to find the least-squares equations for both a linear and an exponential trend line. Complete the calculations, and draw the two trend lines on separate scales (one on an arithmetic scale, and the other on a logarithmic scale—as in Figures 3.12 and 3.13). Also, (a) plot the actual values of the population and log population, and comment on the goodness of fit of the two trend lines, (b) project both trend lines on to 1911 and 1921, and use them

to 'predict' the population of England and Wales in those years. How do these predictions' compare with the actual figures (1911 . . . 36·07 million; 1921 . . . 38·89 million)? Which trend line gives the better predictions? Can you account for the discrepancies?

TABLE 3.7
The population of England and Wales, 1841–1901

	x	x^2	Population (millions) y	xy	$\log y$	$x \log y$
1841	−3	9	15·9	−47·7	1·20	−3·60
1851	−2	4	17·9	−35·8	1·25	−2·50
1861	−1	1	20·1	−20·1	1·30	−1·30
1871	0	0	22·7			
1881	1	1	26·0			
1891	2	4	29·0			
1901	3	9	32·5			

3.9. CALCULATIONS BASED ON EXPONENTIAL MODELS

The last section was concerned with graphing exponential growth. We shall now consider the calculations necessary to make projections, predictions, and post-dictions from exponential trends. A *projection* is simply the continuation of a given trend into the future, and is a purely mathematical operation; a *prediction* involves specifying the trend (which can be no better than an informed and intelligent guess and may later prove to be wide of the mark) and then projecting it into the future; a *post-diction* is a statement about past trends.

3.9a. Projection

The sort of questions we may wish to answer are: if a population of 50 million grows at 2 per cent per annum, (a) what will it be after 10 years, and (b) how long before it doubles itself?

In successive years, it will be 102 per cent of what it was in the previous year. Therefore:

after one year it will be $\quad 50 \times \dfrac{102}{100} = 50 \times 1·02$

after two years, $\quad (50 \times 1·02) \times \dfrac{102}{100} = 50 \times 1·02^2$

after three years, $(50 \times 1·02^2) \times \dfrac{102}{100} = 50 \times 1·02^3$

therefore, after ten years $\quad\quad\quad\quad 50 \times 1·02^{10}$

The calculation is a straightforward use of log tables:

Nos.	Logs
$1{\cdot}02^{10}$	$0{\cdot}0086 \times 10$
	$0{\cdot}0860$
$\underline{50 \times}$	$\underline{1{\cdot}6990\ +}$
$60{\cdot}95$	$1{\cdot}7850$

Therefore, the population, after 10 years of growth at 2 per cent p.a., will have increased from 50 million to 60·95 million.

This example illustrates two rules about the use of logarithms:

(i) $\log XY = \log X + \log Y$
(ii) $\log X^n = n \log X$

thus, $\log (50 \times 1{\cdot}02^{10}) = \log 50 + 10 \log 1{\cdot}02$

Therefore, generalizing the answer to our first question, if a population grows from an initial size of P_0 at an annual rate expressed by r (where r is the *ratio* between the population in one year to the population in the previous year), then the population after n years (P_n) is given by

$$P_n = P_0 \times r^n \qquad\qquad \textit{Formula 3.3}$$

which is generally solved by using logs.

This will help us to answer our second question: how long will it take to double the population growing at 2 per cent p.a.?

Substituting in Formula 3.3, since P_n is to be twice P_0,

$$2P = P \times \left(\frac{102}{100}\right)^n$$

$$\therefore 2 = 1{\cdot}02^n$$

$$\therefore \log 2 = \log 1{\cdot}02^n$$

$$\therefore \log 2 = n \log 1{\cdot}02$$

$$\therefore n = \frac{\log 2}{\log 1{\cdot}02} = \frac{0{\cdot}3010}{0{\cdot}0086} = 35$$

Again, this can be generalized: the doubling time (n years) for a population growing at an annual rate expressed by r, is given by

$$n = \frac{\log 2}{\log r} = \frac{0{\cdot}3010}{\log_{10} r}. \qquad \textit{Formula 3.4}$$

In practice, if the growth rate is between about 1 per cent and 10 per cent p.a. (r between $\dfrac{101}{100}$ and $\dfrac{110}{100}$, a good approximation to doubling time can be obtained more quickly by using 'the rule of 70'—i.e. divide 70 by the percentage annual growth:

2% p.a. growth gives a doubling time of $\dfrac{70}{2}$ = 35 years;

5% p.a. growth gives a doubling time of $\dfrac{70}{5}$ = 14 years.

(It is an arithmetical coincidence that the rule of 70 gives such good approximations.)

The above are very simple projections, based upon exponential trends—i.e. growth (or decline) at a constant, continuous proportional rate. In reality, very few rates of change remain constant for long: they are mostly subject to random and cyclic fluctuations, as described in previous sections. An exponential growth curve is a model—a hypothetical and simplified trend to which some real growth rates may briefly approximate or with which they may be compared. It is possible to make projections from more realistic (and therefore complex) models, which incorporate some anticipated fluctuations—e.g. population projections may take account of the varying numbers in each cohort (age group), and how these are likely to affect birth- and death-rates as the cohorts advance through the age-bands in future years. But such models are beyond the scope of this chapter.

3.9b. Prediction

This is much more difficult, because it involves making assumptions about trends. It is rarely satisfactory to continue past trends into the future, particularly in human affairs, because the conditions which gave rise to the trends are unlikely to remain the same. One way of side-stepping this difficulty is to make several predictions, based upon different explicitly stated assumptions. Another, which is really a variant of the first, is to offer two or three projections and to beg the question of which is most likely to be followed.

For instance, Figure 3.14 shows high, middle, and low projections for world energy consumption. Note that the gap between the projections widens the further they advance into the future, producing a divergent area of uncertainty within which predicted outcomes are likely to occur. On the other hand, firm and fairly precise predictions

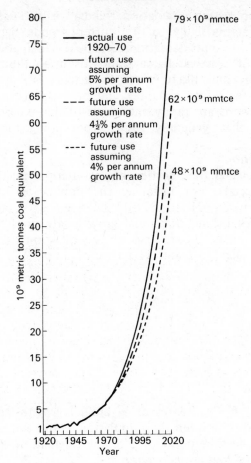

Figure 3.14. High, middle, and low projection for world energy consumption (from O'Dell 1974 p.9)

cannot be avoided by planners, politicians, and entrepreneurs, particularly where decisions have to be made, the effects of which take many years to come to fruition. For instance, prediction of the number of children to be educated in the seventies and eighties determined school-building and teacher-training programmes in the sixties—unfortunately the prediction was wide of the mark, so that school and college buildings, together with thousands of recently trained teachers, are surplus to requirements.

With our present limited understanding of human behaviour, and of some physical processes too, prediction is more of an art than a science. In forecasting population trends, for instance, the art is to identify the factors likely to alter current trends (e.g. the effect of the pill, the economic situation, female emancipation, and rising expectations of material living standards upon future birth-rates), and to quantify their likely effect; the science is then to build these anticipated effects into a model, and project this into the future. Faulty predictions are rarely the result of the experts getting their sums wrong; they generally result from doing the wrong sums.

3.9c. Post-diction

Finally we consider the question of post-diction—making statements about past trends. For instance, the population of Nottingham increased from 28 801 to 57 407 between 1801 and 1851 (see Table 3.2). What kind of annual growth rate does this represent? This is, of course, the reverse of the question asked under the heading of Projection, and is answered in a similar way, using Formula 3.3:

$$P_n = P_0 \times r^n$$

from which $\quad r^n = \dfrac{P_n}{P_0}$

$$r = \sqrt[n]{\dfrac{P_n}{P_0}} \qquad\qquad Formula\ 3.5$$

i.e. r (a ratio P_n/P_{n-1} representing annual growth rate) is the nth root of the ratio between the population at the end and that at the beginning of the time period n. For the calculation, two other rules about the use of logs are necessary:

(i) $\log \dfrac{X}{Y} = \log X - \log Y$

(ii) $\log \sqrt[n]{X} = \dfrac{1}{n} \log X$

Therefore, in this case, with $n = 50$ (years between 1801 and 1851):

	Nos.	Logs
P_n =	57407	4·7590
P_0 =	28801 ÷	4·4594 −
		0·2996 ÷ 50
r =	1·014	0·0060.

Therefore r, the ratio between the population in one year and that of the preceding year, is $1\cdot014$ or $\dfrac{101\cdot4}{100}$, so that the mean annual growth rate of the population of Nottingham between 1801 and 1851 was $1\cdot4$ per cent.

An example will now be given to illustrate a characteristic of ratios which has confused many students. The county reports of the 1971 Census of Population (and of earlier censuses) give the annual inter-censal growth rate of populations broken down into natural increase and net migration, thus (the figures refer to changes in Nottingham C.B. between 1961 and 1971):

Natural Increase %	Net Migration %	Total %
+ 0·68	− 1·11	− 0·37

It is immediately apparent that $0\cdot68 - 1\cdot11$ does not equal $-0\cdot37$. Why not? Briefly, because they are annual rates applied to a ten-year period. To understand this we must consider how the figures were derived. Over the ten-year inter-censal period 1961–71, the proportional changes, all expressed as percentages, were as follows:

(a) natural increase $+ 6\cdot7$
(b) net migration $-10\cdot4$
(c) over-all change $- 3\cdot7$

No problems so far: $6\cdot7 - 10\cdot4 = -3\cdot7$.

Another way of expressing these changes is to say:

(a) if the natural increase had taken place but there had been no migration, the ratio of the 1971 to the 1961 populations would have been $\dfrac{106\cdot7}{100} = 1\cdot067$;

(b) if the migration had taken place without the natural increase, the ratio would have been $\dfrac{100 - 10\cdot4}{100} = \dfrac{89\cdot6}{100} = 0\cdot896$;

(c) in fact, with both natural increase and migration having taken place the ratio was $\dfrac{100 - 3\cdot7}{100} = \dfrac{96\cdot3}{100} = 0\cdot963$.

Now these were the ratios between populations separated by a ten-year period; to get the corresponding *annual* ratios, we use Formula 3.5.

(a) $r_{\text{natural increase}}$ $= \sqrt[10]{1\cdot067} = 1\cdot0068$ (five-figure log tables)

(b) $r_{migration}$ $= \sqrt[10]{0{\cdot}896} = 0{\cdot}9889\,(= 1 - 0{\cdot}0111)$

(c) $r_{nat.\ increase\ and\ migration} = \sqrt[10]{0{\cdot}963} = 0{\cdot}9963\,(= 1 - 0{\cdot}0037)$.

It is from these figures that the annual changes of +0·68 per cent, −1·11 per cent, and −0·37 per cent are obtained.

In short, annual rates of change are obtained from the tenth roots of decennial rates. Now, if A + B = C, then the roots—whether square or tenth roots—of A and B do not add up to the root of C (9 + 16 = 25; 3 + 4 ≠ 5). Hence, while the decennial rates of natural increase and net migration add up to the total rate of population change (because these are the actual decennial rates applied to a ten-year period) the corresponding annual rates do not equate (because they are average annual rates applied to a ten-year period).

Consolidation

6. World population is thought to have reached 4000 million in 1977. Calculate projected populations for the years 2000 and 2030, assuming mean annual growth rates of (i) 2 per cent p.a. (ii) 1·8 per cent p.a. (iii) 1·6 per cent p.a.

7. What were the mean annual growth rates of the population of (i) England and Wales (ii) Nottingham (iii) Ruddington between 1851 and 1871? (See Tables 3.2 and 3.7.)

8. World energy consumption is increasing at about 5 per cent p.a., partly because of increasing world population (2 per cent p.a.). If these trends continued over the next 100 years, by how much would (i) world energy consumption, (ii) world population, (iii) per capita energy consumption then have increased?

4

PROBABILITY DISTRIBUTIONS

4.1. DESCRIPTIVE AND PROBABILITY STATISTICS

THE last three chapters have been concerned with descriptive stat-istics—i.e. with the calculation of indices which summarize data, spatial distributions, and changes through time. But the geographer needs to go further. He may wish to estimate the size, roundness, or composition of the pebbles on a beach, and will want to know how to take a sample, how large the sample needs to be, and how accur-ately he can make his estimates from his sample. He may have sets of data relating to different places—such as land use or crop yields on different slopes or soil formations—and will want to know if the differences or similarities between them are significant or incidental. Or he may wish to find a mathematical relationship or 'model' connecting two or more variables (for instance, the population of a town's 'urban field' and the number and types of shops in its shopping centre), so that he can predict how a change in one variable will affect the others. He will want to draw conclusions from his data. It is generally not possible to do this with complete certainty or precision—geographical problems are too complicated to yield exact solutions. But it is usually possible to calculate the degree of precision or certainty that can be reached from available data. The concepts and techniques required for this purpose can collectively be called 'probability statistics'. and it is with these that most of the rest of this book will be concerned.

The reader will find that some of the material in this chapter seems remote from the problems of geography, but it is hoped that he will not grow impatient on that account. If he is to understand the succeeding chapters, in which techniques for attacking geographi-cal problems are outlined and discussed, he needs first to grasp thoroughly the concepts which are the subject of this chapter.

4.2. PROBABILITY

Probability may be described as the relative frequency with which an event occurs or will occur *in the long run*. It is normally measured on a scale ranging from 0 (meaning impossible) to 1 (meaning certain). For instance, if a 'fair' coin is tossed 100 times, it will turn up heads about 50 times and tails about 50 times. Therefore the probability of it turning up heads is 50 out of 100—i.e. 50/100, or

0·5. The probability of it turning up tails is also 0·5. This can otherwise be expressed as follows:

$$p(\text{heads}) = p(\text{tails}) = 0·5$$

Other examples of probability statements are:

$p(\text{dying}) = 1$ (i.e. 100 out of 100 people will die—death is certain). $p(\text{cat mating with dog}) = 0$ (i.e. a cat has never mated with a dog, is never likely to—it is impossible).

Two further points:

(a)   The knowledge of how often an event will occur in the long run may give little guidance as to when it will occur in the short run. If you have won the toss five times in a row, you are just as likely to win it again next time as if you had lost five times in a row.

(b)   If the probability of an event occurring is known, then the longer the run, the more accurately one can predict the relative frequency or proportion of occurrences. The proportion of heads and tails from 1000 tosses of a coin can be predicted to be near 50:50 with greater confidence than from 100 tosses; and the result of 100 tosses is similarly more predictable than from 10.

Consolidation

1. (a)   If a 'fair dice is thrown 600 times, about how many times will a six be thrown?

   (b)   What is the probability of throwing a six?

2. About 106 boys are born to every 100 girls.

   (a)   What is the probability that the next child born will be a boy?

   (b)   What is the probability that it will be a girl?

   (c)   What is the probability that it will be either a boy or a girl?

4.3. LAW OF ADDITION

If a dice is thrown 600 times, a one will be thrown about 100 times, a two 100 times, and so on. Therefore *either* a one *or* a two will be thrown 200 times. In other words:

$$p(\text{one } or \text{ two}) = 100/600 + 100/600 = 1/6 + 1/6 = 1/3.$$

More generally, the probability of *either* of two independent events occurring is obtained by *adding* the probabilities of the events occurring individually. *Example*: From the study of a farm's records, it is found that over the past 100 years, heavy crop losses were suffered:

in 5 years as the result of spring frost; in four years as the result of summer drought; and in 7 years as the result of autumn storms.

Assuming no over-all change in conditions, the probability of heavy losses from any of these causes is given as follows:

p(spring frost damage) $= 5/100 = 0.05$
p(summer drought damage) $= 4/100 = 0.04$
p(autumn storm damage) $= 7/100 = 0.07$

Therefore, if it is further assumed that there is no connection between the incidence of the three types of weather, p(spring frost *or* summer drought *or* autumn storm damage) $= 0.05 + 0.04 + 0.07 = 0.16$. In other words, heavy crop losses from any one of the above weather types can be expected on about 16 occasions in 100 years, although the data give no indication of when these will occur.

4.4. LAW OF MULTIPLICATION
If two dice are thrown together, there are 36 possible outcomes, which can be tabulated as in Table 4.1:

TABLE 4.1
All 36 possible outcomes from throwing two dice together

6,6	6,5	6,4	6,3	6,2	6,1
5,6	5,5	5,4	5,3	5,2	5,1
4,6	4,5	4,4	4,3	4,2	4,1
3,6	3,5	3,4	3,3	3,2	3,1
2,6	2,5	2,4	2,3	2,2	2,1
1,6	1,5	1,4	1,3	1,2	1,1

Each of these outcomes is equally probable if the two dice are fair. The probability of throwing a double six is therefore $1/36$ or 0.028. This can otherwise be calculated as follows:

p(six) $= \frac{1}{6}$
p(double six) $= \frac{1}{6} \times \frac{1}{6} = \frac{1}{36}$

More generally, the probability of two or more events occurring *together* (either simultaneously or in sequence) is obtained by *multiplying* the probabilities of the events occurring individually. *Example*: Using the data in the example in section 4.3, the probability of heavy crop losses being suffered in two successive years is given by multiplying together the probabilities of heavy losses in each of the years individually.

p(heavy loss in one year) $= 0.16$

Therefore, p(heavy loss in two successive years) $= 0.16 \times 0.16 = 0.0256$. (It is assumed that conditions in one year do not affect conditions in another year).

We can say then that the addition law applies to *either/or* situations and the multiplication law to *and* situations.

Consolidation
3. If two dice are thrown together, what is the probability of a double being thrown?
4. What is the probability of throwing two dice so that the sum of their score is (a) 7, (b) 10, (c) more than 8?

4.5. PROBABILITY DISTRIBUTIONS

Look again at the table of possible outcomes of throwing two dice in the previous section. If the pairs of numbers are summed, the result is as in Table 4.2.

TABLE 4.2
The sums of all possible outcomes of throwing two dice

12	11	10	9	8	7
11	10	9	8	7	6
10	9	8	7	6	5
9	8	7	6	5	4
8	7	6	5	4	3
7	6	5	4	3	2

We can see that there is only one way of getting a score of 12, two of getting 11 (a 5 and 6 or a 6 and 5), three of getting 10, and so on. Now it will be recalled that each 'way' (i.e. outcome of throwing two dice) has an equal probability of occurring—$\frac{1}{36}$. Therefore the probability of getting a score of 12 is $\frac{1}{36}$, of getting 11 is $\frac{2}{36}$, of getting 10 is $\frac{3}{36}$, and so on. (This is applying the addition law.) A full table of all the possible scores and their associated probabilities can now be made:

Score:	12	11	10	9	8	7	6	5	4	3	2
Probability:	$\frac{1}{36}$	$\frac{2}{36}$	$\frac{3}{36}$	$\frac{4}{36}$	$\frac{5}{36}$	$\frac{6}{36}$	$\frac{5}{36}$	$\frac{4}{36}$	$\frac{3}{36}$	$\frac{2}{36}$	$\frac{1}{36}$

A distribution of all possible outcomes, together with their associated probabilities, such as this, is known as a 'probability distribution'. It will be seen that if all the probabilities are added together, they make 1, which means that it is certain that the score from throwing two dice will be a whole number between 2 and 12 inclusive.

The probabilities in a probability distribution add up to 1, because the distribution covers all possible outcomes.

4.6. THE BINOMIAL DISTRIBUTION

One of the commonest probability distributions is associated with the repetition of events where there are only two possible outcomes. For instance, when a baby is born, it is either a boy or a girl. The probabilities are about equal (0·5 each—we shall ignore the slightly larger number of boys born, which was noted earlier in the chapter). To make the example more general, however, we shall call the probability of a boy being born p, and the probability of a girl being born q. Then $p + q = 1$, as these are the only possible outcomes of a birth.

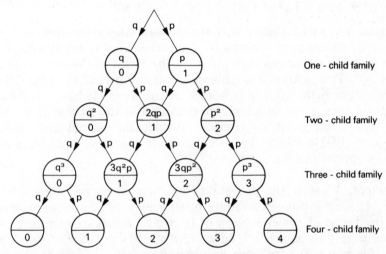

Figure 4.1. Probabilities associated with the number of boys in families of different size

Consider now the probabilities of having different numbers of boys in one-, two-, three-, and four-child families. Figure 4.1 sets these out. Each horizontal line of circles shows all the possibilities for a family with the stated number of children. The term in the upper half of each circle is the probability of having the number of boys shown in the lower half of the circle.

Thus, in a three-child family, the probability of having two boys is $3qp^2$. The reason is suggested by the diagram. To get to this circle requires two 'moves' (i.e. births, shown by lines connecting the circles) of probability p each, and one of probability q. The probability of doing this *once* is $q \times p \times p = qp^2$ (multiplication law). But there are three possible ways of doing this ($p\,p\,q, p\,q\,p, q\,p\,p$)

—i.e., there are three alternative routes to the circle in question (boy, boy, girl; boy, girl, boy; girl, boy, boy). The total probability of reaching this circle is therefore $qp^2 + qp^2 + qp^2 = 3qp^2$ (addition law).

Consolidation
5. Copy Figure 4.1, and complete the bottom line of circles.

It will be observed that the terms in the upper halves of each line of circles are the terms of a familiar algebraic expansion:

$$
\begin{aligned}
q + p &= (q + p)^1 \\
q^2 + 2qp + p^2 &= (q + p)^2 \\
q^3 + 3q^2p + 3qp^2 + p^3 &= (q + p)^3 \\
q^4 + 4q^3p + 6q^2p^2 + 4qp^3 + p^4 &= (q + p)^4
\end{aligned}
$$

In fact, it can be shown that the probabilities associated with all possible outcomes of n births (or any other kind of event with only two possible outcomes) are given by the terms in the expansion of $(q + p)^n$. This is known as a binomial expansion, and the probability distribution is therefore a 'binomial distribution'. It has a very wide range of uses, some of which are discussed in the next chapter, but for the moment let us consider further the special case, where $p = q = 0.5$ (to which the question about the number of boys in a family approximates).

Example: What is the probability of there being no boys, 1 boy, 2, 3, 4, 5, or 6 boys in a family of six children? The answer is given by the terms of the expansion of $(q + p)^6$, which are (omitting the connecting plus signs): $q^6, 6q^5p, 15q^4p^2, 20q^3p^3, 15q^2p^4, 6qp^5, p^6$. But if $p = q = 0.5$, this can be restated as follows (with the corresponding number of boys written below).

$(0.5)^6$	$6 \times (0.5)^6$	$15(0.5)^6$	$20(0.5)^6$	$15(0.5)^6$	$6(0.5)^6$	$(0.5)^6$
0 boy	1 boy	2 boys	3 boys	4 boys	5 boys	6 boys

Since

$$(0.5)^6 = \frac{1}{2^6} = \frac{1}{64},$$

these probabilities can be written as:

$$\frac{1}{64} \quad \frac{6}{64} \quad \frac{15}{64} \quad \frac{20}{64} \quad \frac{15}{64} \quad \frac{6}{64} \quad \frac{1}{64}$$

or   0·016   0·094   0·234   0·313   0·234   0·094   0·016

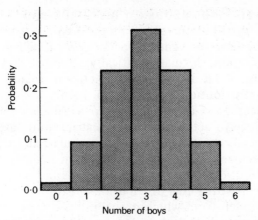

Figure 4.2. Probabilities associated with different numbers of boys in a six-child family

This result can be shown by a histogram, as in Figure 4.2.

It will be noticed that:

(i) The distribution is symmetrical—there is just as much likeli-hood of having six boys (and no girls) as there is of having no boys (and six girls). This is because $p = q = 0.5$. Only when this is so is a binomial distribution symmetrical.

(ii) The mean is three—that is, if one divided the number of boys in six-child families by the total number of such families, one would expect an answer of three.

(iii) The greatest probabilities cluster around the mean—most six-child families will have about half boys and half girls.

(iv) The greater the deviation from the mean, the smaller the probability of that number occurring. It is very unlikely that all six children will be boys, or all six girls.

This is of no particular help to parents with five children, all girls, who are wondering whether to 'try once more' for a boy! The pro-bability that the next one will be a boy is still about 0.5 (unless there is some genetic bias to produce one sex rather than the other). But assuming there is no genetic bias, it does mean that for every 1000 six-child families, about 16 will have six girls and no boys, 94 will have five girls and a boy, 234 will have four girls and two boys, 313 will have three girls and three boys, 234 will have two girls and four boys, 94 will have one girl and five boys, and 16 will have six boys and no girl. If genetic bias is suspected, a binomial distribution would provide the norm against which to compare frequencies from a random sample of families.

4.7. PROBABILITY AND FREQUENCY DISTRIBUTIONS

A probability distribution represents the frequency with which one would expect values (or events) to occur *in the long run* — it is a hypothetical distribution, is mathematically derived, and is perfectly regular. The binomial distribution depicted in Figure 4.2 is such a hypothetical distribution: one would hardly expect 50 or 100 randomly selected six-child families to yield such perfectly symmetrical results. An actual frequency distribution, based on a finite number of occurrences, is likely to be irregular; and the smaller the number of occurrences, the more irregular it is likely to be. A probability distribution 'irons out' the incidental irregularities that inevitably occur with actual distributions. It is therefore simpler, and serves as a 'model from which probabilities can be estimated more easily and more reliably than from a distribution of actual frequencies.

4.8. THE NORMAL PROBABILITY DISTRIBUTION

In this and the following two sections we shall discuss the most important probability distribution in statistics — the normal distribution. It is a hypothetical distribution to which a large number of actual distributions approximate. Let us begin by looking at one such actual distribution.

Table 4.3 sets out the annual rainfall figures, correct to one decimal place, for Derby Sewage Works (hereafter simply referred to as Derby), for the 50 years 1917 to 1966 inclusive:

TABLE 4.3
Annual rainfall at Derby Sewage Works 1917—1966 inches

1917—9	1920—9	1930—9	1940—9	1950—9	1960—6
	27·5	37·0	24·1	24·1	37·0
	19·6	30·1	24·8	27·1	22·4
	27·6	28·8	19·2	23·5	19·7
	25·7	21·5	20·2	21·4	22·4
	28·6	20·7	24·0*	28·4	19·8
	30·3	27·6	21·7	21·2	31·8
	23·7	24·1	27·5	24·5	33·6
24·2	30·9	23·3	21·3	25·8	
25·8	28·8	21·6	24·9	29·3	
27·9	24·2	24·7	22·2	17·8	

*In fact, 23·98, and therefore counted in group 22—4 below.

These figures can be put into the form of a grouped frequency distribution, as follows:

Annual rainfall	16–18	18–20	20–2	22–4	24–6	26–8	28–30	30–2	32–4	34–6	36–8
Frequency	1	4	8	7	12	6	5	4	1	0	2

This frequency distribution can be illustrated by a histogram, as in Figure 4.3.

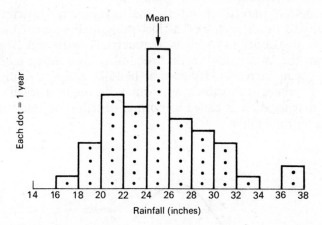

Figure 4.3. Grouped frequency distribution of annual rainfall at Derby Sewage Works, 1917 to 1966

The mean and standard deviation have been calculated to be:

Mean: 25·3 inches
Standard deviation: 4·3 inches.

As this is an actual distribution, we are not surprised that it does not possess the regularity of the binomial distribution shown in Figure 4.2. The question arises: is there a regular, hypothetical frequency distribution—i.e. a probability distribution—to which this would approximate *in the long run* (i.e. with a much longer run of years, assuming no long-term climatic change)? Because, if there is, and if its mathematical properties can be determined, it will provide a means of calculating rainfall probabilities for Derby.

There are both empirical and logical reasons for believing that there is. Empirically, the shape of the histogram (Figure 4.3), while not perfectly regular, is by no means random and shows some symmetry: most of the values are clustered near the mean, and frequencies decline about equally in both directions with increasing distance from the mean. Rainfall data for other stations with widely differing regimes have similar characteristics (except for stations with very low rainfalls). Logically, one would expect such a pattern.

Rainfall appears to be affected by two groups of factors: one dominant group that acts consistently and causes clustering about the
mean (e.g. latitude, altitude, and exposure to sea influence); and
another group of factors that acts randomly, and causes, in the long
run, a symmetrical variation about the mean (e.g. the variable paths
followed by depressions and anticyclones).

It is generally agreed, therefore, that a frequency distribution of
annual rainfall totals such as that just presented has a corresponding
probability distribution which is symmetrical, with highest probabilities near the mean, and with probability decreasing as distance
from the mean increases. Its general bell-like shape is indicated in
Figure 4.4, where it is shown by a curve, in relation to the rainfall
data just discussed. It is called a 'normal distribution' and the curve
is called a 'normal curve'.

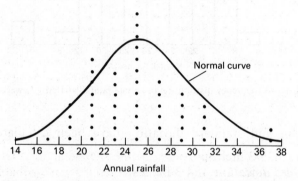

Figure 4.4. Frequency distribution of annual rainfall at Derby related to a normal curve

Before going on to discuss in detail the characteristics and properties of the normal distribution, it may be said that whenever there
appear to be two groups of factors affecting a variable—one dominant group acting consistently, and another acting randomly—then
there are *a priori* grounds for expecting that a frequency distribution
of values will, in the long run, approximate to a normal distribution.
The roundness of pebbles on a beach; the height of remnants of an
unwarped and untilted erosion surface; temperature and other
weather data; crop yields under uniform physical conditions; pedestrian and traffic densities for particular places at particular times of
the week; population data for towns such as average age, percentage
employed in distributive trades, or rates of population growth; these
are just a few variables that are likely to show a tendency towards a
normal distribution.

Question for discussion: What are the 'dominant factors acting consistently' and 'factors acting randomly' in the above examples? Can you think of other examples?

4.9. PROPERTIES OF THE NORMAL CURVE

Let us now examine the normal distribution in more detail. First, notice the similarity in general shape of the binomial (where $p = q$) and normal distributions: in both, probabilities fall away symmetrically from peaks at the mean. But, whereas the normal distribution is represented by a curve, the binomial is represented by a histogram. This is because the normal distribution is related to values on a continuous scale. Rainfall can have any value, whereas the number of boys in a family must be a whole number—you can have 4·75 inches of rain, but not 4·75 boys. Similarly, a normal curve gives probabilities for any range of values, while a binomial can only do so for discrete values such as whole numbers.

Implicit in the normal curve being related to a continuous scale is the fact that it is related to an infinite number of possible values—for instance, 25, 25·1, 25·01, 25·001, etc. The probability of any particular value occurring is therefore infinitesimally small, and can be regarded as zero. One can only speak of the probability of values occurring within a given *range*, and this is shown by the *area* under the appropriate part of the curve, and not by its height. The point is illustrated in Figure 4.5, where the probability of values lying between x_1 and x_2 is proportional to the shaded area under the curve. Furthermore, since the normal curve describes a probability distribution, the total area under the curve is proportional to the total probability (1).

As the normal curve is symmetrical about the mean, half the area beneath the curve lies on each side of the mean. This implies that, in

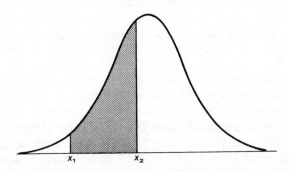

Figure 4.5. The area under a normal curve as a measure of probability

a frequency distribution which approximates to normal, 50 per cent of the values will be less than the mean, and 50 per cent more than the mean. (On a continuous scale, it is highly unlikely that any value will be *exactly* on the mean—it will almost certainly be at least fractionally either above or below.) In other words, the probability of a value being less than the mean is equal to the probability of it being greater than the mean, and both are equal to 0·5.

(*N.B.* Relative frequencies are usually expressed on a percentage scale, while probabilities are generally expressed on a 0 to 1 scale. But as probability can be regarded as 'frequency in the long run', it is also sometimes expressed on a percentage scale—e.g. on probability paper, see p.128).

Less obvious properties of the normal curve, arising from the mathematical equation determining its shape (a very complicated equation which need not detain us), relate to its standard deviation. It is found that approximately:

> 68 per cent of the area under a normal curve lies within one standard deviation of the mean;
> 95 per cent of the area lies within two standard deviations of the mean;
> 99·7 per cent of the area lies within three standard deviations of the mean.

These facts are illustrated in Figure 4.6, which shows normal curves with mean μ and standard deviation σ.

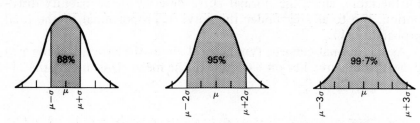

Figure 4.6. The area under a normal curve within one, two, and three standard deviations of the mean

With this information, it is possible to estimate certain probabilities about any variable whose distribution is approximately normal, if we know the mean and standard deviation. For instance, we have seen that the distribution of annual rainfall totals is approximately normal, and that on the basis of data already quoted, the mean annual rainfall of Derby is 25·3 inches, with a standard deviation of 4·3 inches. We can therefore make the following probability statements about Derby's annual rainfall (R):

(i) the probability that R will be less than $25 \cdot 3 = 0 \cdot 5$
     i.e. $p(R < 25 \cdot 3) = 0 \cdot 5$

Equally

(ii) $p(R > 25 \cdot 3) = 0 \cdot 5$

Also,

(iii) the probability that R lies between 21 inches and $29 \cdot 6$ inches
       is $0 \cdot 68$. (21 in. is 1 standard deviation below the mean, and
       $29 \cdot 6$ in. is 1 S.D. above the mean.)
       This can otherwise be stated: $p(21 < R < 29 \cdot 6) = 0 \cdot 68$.

Similarly

(iv) $p(16 \cdot 7 < R < 33 \cdot 9) = 0 \cdot 95$ (two S.D.s below and above the
      mean)

and

(v) $p(12 \cdot 4 < R < 38 \cdot 2) = 0 \cdot 997$ (three S.D.s below and above
     the mean).

From these statements, we can deduce the number of occasions in
50 years when we would expect the annual rainfall to be within the
stated ranges, and it is interesting to compare these expected frequencies with the actual frequencies between 1917 and 1966:

	Expected	Actual
R less than 25·3 in.	25	29
R more than 25·3 in.	25	21
R between 21 and 29·6 in.	34	36
R between 16·7 and 33·9 in.	47 or 48	48
R between 12·4 and 38·2 in.	50	50

The discrepancies between the two sets of figures could simply be
due to the expected frequencies being derived from a regular probability distribution (i.e. the normal distribution), while the actual
frequencies contain the usual incidental irregularities.

Consolidation
In Figure 4.7, a, b, and c represent the probability of any normally
distributed variable lying between the values indicated. Thus:

$$a + b + c = 0 \cdot 5$$
$$c + c \quad\ = 0 \cdot 68$$
$$2b + 2c = 0 \cdot 95$$

7. What are the values of a, b, and c?

8. What is the probability of x being less than:

    (i) $\mu - 2\sigma$ (ii) $\mu - \sigma$ (iii) μ (iv) $\mu + \sigma$ (v) $\mu + 2\sigma$

9. What is the probability of x being more than:

    (i) $\mu - 2\sigma$ (ii) $\mu - \sigma$ (iii) μ (iv) $\mu + \sigma$ (v) $\mu + 2\sigma$

10. The 'A' horizon of a podzolic soil in a locality of generally uniform terrain has a mean depth of 10 cm, with a standard deviation of 2 cm Assuming the depth of the 'A' horizon to be normally distributed, what is the probability that a randomly selected soil profile will have an 'A' horizon of the following depths:

    (i) less than 8 cm;   (ii) more than 14 cm;

   (iii) between 10 and 12 cm;   (iv) exactly 10 cm;

    (v) between 10 and 14 cm?

Let us return again to the rainfall at Derby. Suppose we wish to know the probability of the annual rainfall being less than 20 inches.

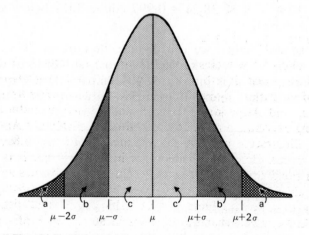

Figure 4.7. The area under different portions of a normal curve

We know that the mean annual rainfall is 25·3 inches and the standard deviation is 4·3 inches; so that 20 inches is not a whole number of standard deviations from the mean. It is 5·3 inches from the mean, which is more than one S.D. but less than two S.D.s from the mean. To solve our problem, we need to know probabilities corresponding to values *between* whole standard deviations from the mean. For this, two steps are necessary:

1. Convert the value in question (in this case, 20 inches) into a 'z-score' (already introduced in Chapter 1, p.43). This is simply the

number of standard deviations of a value above or below the mean. Thus, if the value in question is x, then

$$z = \frac{x - \text{mean}}{\text{standard deviation}} \qquad\qquad \textit{Formula 4.1}$$

This may be written as

either: $z = \dfrac{x - \mu}{\sigma}$,   where μ and σ are the mean and standard deviation of a population.

$$\textit{Formula 4.1a}$$

or: $z = \dfrac{x - \bar{x}}{s}$,   where $\bar{x}$ and s are the mean and standard deviation of a sample.

$$\textit{Formula 4.1b}$$

In the problem before us:

$x = 20$ in.;   mean $= 25 \cdot 3$ in.;   standard deviation $= 4 \cdot 3$ in.

Therefore:

$$z = (20 - 25 \cdot 3)/4 \cdot 3 = -1 \cdot 2.$$

2. Consult one of the tables of probabilities associated with different z-scores in a normal distribution. Appendix 1 col. A gives the probabilities of a variate lying *between* the mean and corresponding values of z, and Appendix 1 cols. B and C give probabilities of a variate lying *beyond* corresponding z-values. A p-value in Appendix 1 col A is illustrated by the shaded area in Figure 4.8a; that in Appendix 1 col. B by the shaded area in Figure 4.8b. It is evident, both from the diagrams and from the tables, that p-values *within* and *beyond* a given z-value add up to $0 \cdot 5$, because this constitutes the probability of a variate lying in one half of a normal distribution.

It should be noted that the p-values in these two tables relate to only half the normal distribution; that one half of a normal distri-

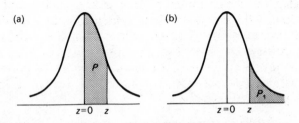

Figure 4.8. (a) and (b) Areas under normal curve representing probabilities associated with a given z-score

bution is a mirror-image of the other; and that therefore a negative
z-score has the same p-value as a positive z-score of equal magnitude,
so that we can ignore the sign of z when using the tables.

In our problem, z is 1·2, and we wish to know the probability of a
variate being more extreme than—i.e. lying *beyond*—this z-score.
We therefore use Appendix 1 col. B, where the value of p corresponding to z = 1·2 is 0·115. It follows that the probability of the
rainfall being less than 20 inches is 0·11, meaning that an annual
rainfall of under 20 inches can be expected at Derby eleven years in
every hundred.

Consolidation
11. (i) What is the probability that the rainfall at Derby will be at
least 28 inches?
    (ii) What is the probability that it will be between 20 inches and
30 inches?

4.10. THE CUMULATIVE PROBABILITY CURVE OF A NORMAL DISTRI-BUTION

A cumulative probability curve shows the probability of a value
being less than a specified amount. Just as an ordinary grouped frequency distribution can be depicted by a cumulative frequency curve
(see Chapter One, p.9), so a normal probability distribution has a
corresponding cumulative probability curve. Its general shape can be
seen in Figure 4.9, which is based on the following data, which
(should) have been worked out in Consolidation Exercise 8.

Value less than:	$\mu - 2\sigma$	$\mu - \sigma$	μ	$\mu + \sigma$	$\mu + 2\sigma$
Probability:	0·025	0·16	0·5	0·84	0·975

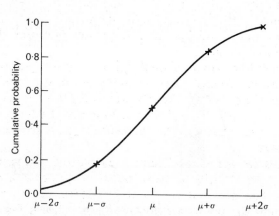

Figure 4.9. Cumulative probability curve for a normal distribution

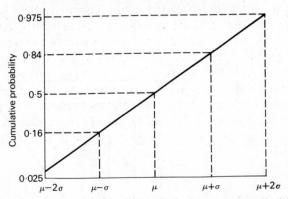

Figure 4.10. Straightened cumulative probability curve for a normal distribution

By adjusting the vertical scale, the S-shaped curve of Figure 4.9 can be turned into the straight line of Figure 4.10. Graph paper with the vertical scale so adjusted is known as 'probability paper' (see Figure 4.11, and note that here probabilities are measured in percentages). A cumulative probability graph of a normal distribution is a straight line when drawn on *probability paper*. This may be compared with an exponential growth graph, which becomes a straight line on semilogarithmic graph paper (see Chapter Three, p.102).

Probability paper therefore has the following uses:

1. To check whether or not an actual frequency distribution is approximately normal; and if it is:
2. To estimate the corresponding probability distribution;
3. To read off its standard deviation and mean;
4. To read off any required probabilities.

The results obtained by graphical methods are somewhat inexact when compared with the results of mathematical calculation, but they are very much quicker, and for many purposes the level of precision is quite adequate.

How probability paper is used for the above purposes is best explained with an example, again making use of the rainfall data for Derby. These will be restated in two ways (Table 4.4): (a) in grouped frequencies (as on p.119); and (b) in cumulative frequencies and percentages—i.e. the number and percentage of occurrences below different values.

The cumulative percentages have been plotted in Figure 4.11, on probability paper. It will be seen that they do not exactly lie on a straight line. The distribution is not, therefore, exactly normal. On the other hand, the points are spread out in a linear fashion, which means they represent a distribution which is approximately normal.

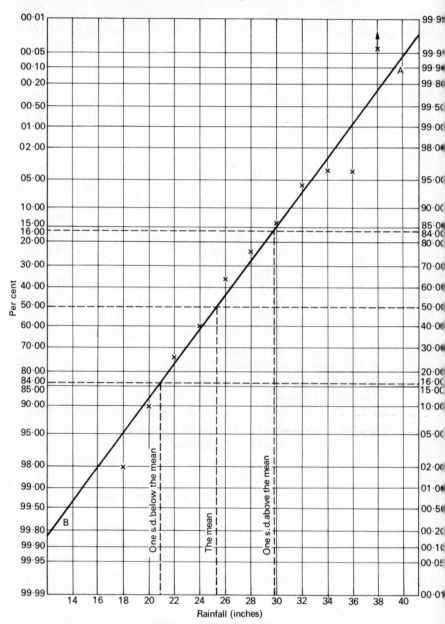

Figure 4.11. Annual rainfall figures for Derby plotted on probability paper

It is possible to estimate a corresponding normal probability distribution by drawing a best fit' line (AB) through the points.

We can now read off:

(a) The mean and standard deviation of the normal probability

TABLE 4.4
*Grouped and cumulative frequency data for annual rainfall at Derby
1917–1966*

(a) *Grouped frequencies:*

Annual rainfall	16–18	18–20	20–2	22–4	24–6	26–8	28–30	30–2	32–4	34–6	36–8
Frequency	1	4	8	7	12	6	5	4	1	0	2

(b) *Cumulative frequencies:*

Rainfall below	18	20	22	24	26	28	30	32	34	36	38
Cu. Freq.:	1	5	13	20	32	38	43	47	48	48	50
Cu. %	2	10	26	40	64	76	86	94	96	96	100

distribution to which the actual distribution of rainfall values
approximates. The mean is the value on the horizontal scale corresponding to the 50 per cent level on the vertical scales. (The two
vertical scales are complementary: that on the left shows the percentage probability of being *above* corresponding values on the
horizontal scale; that on the right shows the probability of being
below those values.) It is seen to be about 25·3 inches. The standard
deviation is obtained from the difference between the mean and the
horizontal value corresponding to either the 16 per cent or 84 per
cent levels on either of the vertical scales. (If the reason for this is
not clear, re-examine Figure 4.10, and the data on which it is based.)
It is seen to be very close to the calculated value of 4·3 inches.

(b) Required probabilities: for instance, the probability of the
rainfall being under 34 inches is just over 97 per cent (right-hand
scale), and of being over 34 inches is just under 3 per cent (left-hand
scale). This can otherwise be stated as follows:

$$p(R < 34 \text{ in.}) = 0·97 \quad \text{and} \quad p(R > 34 \text{ in.}) = 0·03.$$

Consolidation
12. Using the graph in Figure 4.11, determine the following probabilities:
(a) that Derby's rainfall in a single year will: (i) exceed 35 inches,
     (ii) be at least 32 inches, (iii) be less than 25 inches, (iv) be be-
     tween 25 and 30 inches.
(b) that the rainfall will (i) in two successive years be less than 25
     inches, (ii) in three successive years be over 30 inches.

Also determine the minimum and maximum amounts of rainfall to
be expected in 9 years out of 10.

5

SAMPLES AND ESTIMATES

5.1. INTRODUCTION

SAMPLING has been used in two ways by geographers: as an expository technique (the 'sample study' method of teaching), and as a means of investigation. Furthermore, samples may be used in investigations either to gain a general impression of an area, or as the basis for making estimates of some of its numerical characteristics, in which case they are sometimes distinguished as 'statistical' samples. This chapter is only concerned with this last sense of the term, and the word sample will henceforth carry this statistical connotation.

5.2. POPULATIONS AND SAMPLES

In statistics the word *population* has a technical meaning which may have nothing to do with people. Suppose we wish to determine various aspects of the field pattern over a specified tract of land—e.g. size and compaction of fields, type of boundary, land use. If each of these characteristics were noted for every field, we should have four populations: a set each of size and compaction *measurements* would constitute two populations; and sets of *counts* of different types of boundary and land use would constitute the other two. In statistics, *a population is a complete set of counts or measurements derived from all objects possessing one or more common characteristics*—in the foregoing example the common characteristics being designated by the words 'field' and 'over a specified tract of land'.

A population may have a finite number of individual counts or measurements (e.g. the sizes of factories in a town); it may be almost infinite (e.g. the composition of pebbles on a beach); or it may be infinite (e.g. the slope of the land at point locations on a map, there being an infinite number of points on a map).

It is often neither necessary nor practicable to obtain all the counts or measurements for a complete population. Suppose one wished to know the average field-size or the proportion of stone-wall field boundaries over our tract of land, and suppose there were several thousand fields involved. One could obtain fairly good, and for most purposes perfectly adequate, approximations by measuring or counting only a fraction of all the fields, and thereby save a great deal of time and trouble. The data so collected would constitute *a sample* which may be defined simply as part of a population. Note that in statistics a sample, like a population, is a collection of data, not of things.

(It should be said, however, that the terms 'population' and 'sample' as used in this and other books, often appear to refer to the objects under study rather than to data relating to those objects. This is largely for the sake of brevity—it is easier to say that 'the population consists of all households' than 'the population consists of specified data relating to all households'.)

A sample is, therefore, used as a substitute for population, to save time or where measuring or counting the whole population is impractical. If it is to be a reliable substitute, it must be a *representative sample*—i.e. it must represent the variability in the population as closely as possible. The proportion of large, medium, and small fields in the sample must be about the same as in the population. To achieve this, it is important to avoid *bias*, which results in the inclusion in the sample of unrepresentative proportions of data from different parts of the population.

If samples are selected entirely by personal judgement on the part of the sampler, bias is almost bound to occur. Subjective selection cannot free itself from subconscious predilictions of the sampler for picking up small pebbles rather than large, observing plants with striking colours more often than those of a more subdued hue, or interviewing attractive members of the opposite sex. And even if subjective selection is made without bias, this cannot be demonstrated and there can be little confidence that the sample is truly representative. It is essential, therefore, that, if the sample is both to be. and to be *seen* to be without bias, the selection of individuals should not depend upon the judgement of the sampler. Methodical procedures must be used, which reduce the subjective element in selection to a minimum.

5.3. THE SAMPLING FRAME

Where judgement is required, however, is in the choice of suitable sampling frame—that is, of a statistical population from which to take the sample. For often the true population is not amenable to sampling in the time available, and a substitute has to be found. In the 10 per cent Sample Census of 1966, it was impractical to question 10 per cent of all individual people. Instead, 10 per cent of households were sampled. All the people of Britain constituted the true population, but it was the households of Britain which constituted the sampling frame from which a 10 per cent sample was taken.

The sampling frame may be a serious source of bias, depending on how representative it is of the true population being studied. A telephone directory would provide a very good sampling frame if the true population were all a town's telephone subscribers, or its business concerns (nearly all of which can be assumed to be on the 'phone); it would be overweighted with relative affluence if it were used to represent all the town's householders. Weekday shoppers would also provide a poor sampling frame from which to draw a sample of public opinion: it would contain too many housewives and not enough people in employment. Transects radiating from the centre of a city may provide a useful sampling frame for estimating the relationship between distance from the centre and some other variable; it will over-represent areas near the city if used to estimate the proportion of land given over to different uses (because the space between radiating lines increases with distance from the centre). Radial transects provide a good sampling frame when the true population under consideration is 'all possible intra-city distances from the centre', they do not if the true population is 'all possible locations within the city'.

Bias may occur in more subtle ways. A good example is given in Gregory (1963), p.94. If one is choosing a sample of farms from a gridded map, the sampling frame may be *either* (i) farms on whose land selected grid intersections occur, *or* (ii) farms whose farmhouses are nearest the selected intersections. Gregory comments: 'the former will tend to give an over-representation of the number of large farms, but a true representation of the amount of land held by large farms; the latter will tend to over-represent the small farm in terms of the amount of land that falls under small farms, but to give a true representation of the number of small farms.' In such a case, the choice of sampling frame depends upon a very precise definition of the true population.

Sampling frames may be either spatial or non-spatial. A *spatial sampling frame* is one in which location is an essential part of the variability of individual observations, and where it is therefore important that samples are representative of the full range of locational variability. If, for example, one is sampling a map to determine the amount of woodland in the map area, one must ensure that all parts of the map are duly represented. This may be done in one of three ways:

    (i)   Point-sampling—e.g. by sample grid-intersections;
    (ii) Line-sampling—e.g. by a sample of linear transects or cross-sections;
    (iii) Area-sampling—e.g. by a sample of quadrats or grid-squares.

The sampling frames in these cases will be the populations of all possible grid-intersections, linear transects, and quadrats on the map respectively.

Although geographers are particularly interested in spatial sampling frames, they also make frequent use of the non-spatial variety, where the locational variability of individual observations is not considered. The sampling frame is then simply the total collection of phenomena under consideration, such as all the households in the parish, all the shops in the Central Business District (C.B.D.), or all the shoppers in the High Street. It is still important to sample the full range of variability within the population, but this can be done without reference to their location—e.g. by using an electoral register or telephone directory, or by simply stopping people outside Woolworth's.

Whether one uses a spatial or non-spatial sampling frame depends upon the object of the inquiry. If it is to find out something non-spatial about an area *as a whole* (such as the number of high-order shops in a city's C.B.D.—assuming the C.B.D. is already defined) then a non-spatial sampling frame may be adequate. If it is to find out something spatial about an area as a whole (such as the amount of land occupied by high-order shops in the C.B.D.), or if it is to make spatial distinctions within the area (such as differences in rateable value between the C.B.D.'s core and frame), then a spatial sampling frame is required.

5.4. SAMPLING PROCEDURES

Having decided upon a suitable sampling frame, it is then necessary to select individuals for the sample by a method which is as mechanical as possible.

The two most common sampling methods are (i) random sampling, and (ii) systematic sampling.

5.4a. *Random sampling*

A random sample has been defined as one 'in which any one individual measurement or count in the population is as likely to be included as any other' (Alder and Roessler, 1964). The usual way of obtaining a random sample is by the use of a table of random numbers. The use of random numbers is best explained by an example.

Suppose we wish to obtain a random sample of 20 farms from the area shown in Figure 5.1, with a view to investigating various aspects of the farm economy of the area. One method might be as follows:

(i) Begin by choosing a sampling frame which will ensure a random spacing of farms. This is often done by using the intersections

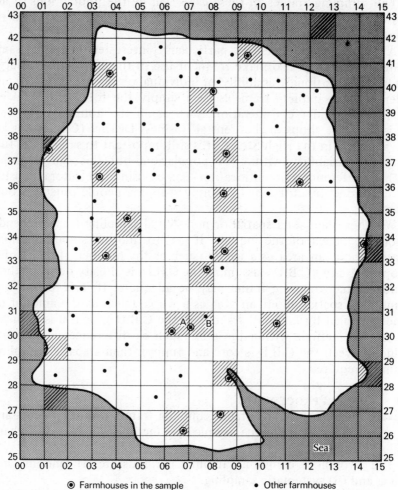

Figure 5.1. Farmhouses in a random sample of 25 grid squares

or squares of a superimposed grid. Let the sampling frame in this case be all farms with their farmhouse nearest the south-west corner of the grid square in which they are located'. This means that we first require a random sample of grid squares, so that we can select the farm nearest the south-west corner in each case. Because some of the squares may contain no farm, we shall obtain a random sample of 25 squares, and then choose the first 20 that contain farmhouses.

(ii) To obtain the sample of grid squares turn to the table of random numbers (Appendix 2). Select a pair of digits from anywhere in the table, and work methodically across the table in any direction, reading off two digits at a time. We shall start at the bottom right-hand corner of the table, and work up the right-hand column to the top, then down the next column, up the next again, and so on.

(iii) The numbers are first used to determine the eastings of the sample grid squares. These can only run from 00 to 14, so we shall ignore most of the pairs of random numbers we read off. The first 25 pairs of digits that come within our range of eastings are set out in the left hand column of Table 5.1.

TABLE 5.1
*Twenty-five
randomly selected
grid references*

11	36
01	27
07	39
09	41
11	31
10	30
14	28
08	37
12	42
08	28
08	26
01	29
03	40
04	34
08	33
06	30
14	33
03	33
08	35
07	30
07	32
03	36
01	37
06	26
00	30

(iv) When sufficient eastings have been obtained, the reading off of random numbers is continued to obtain the same number of northings within the grid's range of 25 to 42 (right-hand column of Table 5.1)

(v) We now have 25 four-figure grid references, beginning with 1136, and these determine the grid squares in which our sample farms are located (the shaded squares in Figure 5.1). It will be seen that there are five shaded squares without farmhouses, so we have just enough squares from which to select a sample of 20 farms. In the case of those squares containing two or more farmhouses, we select only the nearest farmhouse to the south-west corner of the square.

(vi) It may be necessary to introduce additional rules to avoid having to make judgements about the selection of individual cases. For instance, A lies on a line between two squares. For such cases, a rule is made to assign the farm to the square to the north and/or east of the line. A therefore belongs to square 0730, and, as it is nearer to the south-west corner of that square than B, it is selected for the sample and B is not.

It will be seen that the rules are quite arbitrary. This does not matter, so long as they are simple, do not create bias, and are applied consistently, so that the subjective element in the selection of individual cases is eliminated.

A last point to note is the choice of 'farms in grid-squares' rather than farms nearest grid intersections' as the sampling frame. The reason for this is that coastal farms might otherwise be either under-represented (if grid intersections falling in the sea were ignored) or over-represented (if such grid intersections were used).

Consolidation

1. Use a one-inch O.S. map to estimate the percentage of the map area devoted to different forms of land use (e.g. woodland, water, buildings, routeways, etc.) by random point sampling. This can be done by noting the land use at 100 randomly selected grid intersections.

5.4b. Systematic sampling:

A systematic sample is one in which selection is made from the population at regular intervals. Every tenth name on an electoral list, every fifth shop listed in a classified telephone directory, every alternate grid intersection: these will all yield systematic samples. It is generally quicker and simpler to obtain a systematic sample than a random sample, for random numbers are not required. A systematic sample also gives a more uniform cover of the population, for random samples tend to include clusters and leave gaps unless the random sample is very large. Thus, if one were estimating the proportion of woodland in the map area of Figure 5.2a, and if the major area of woodland happened to be in the vicinity of A, the estimate would be above the true value (other things being equal).

On the other hand, if there are regularities in the background population, a systematic sample may be biased. If one were to impose a rectangular grid over many American cities, a systematic sample based on the grid intersections could consist of data derived mainly from street-corner locations. More generally, a systematic sample is likely to be biased unless the background population from which it is taken is itself randomly distributed—and many background populations are not randomly distributed, as Von Thünen and Christaller (among others) have demonstrated. In fact, the statistical techniques applied to samples, which are discussed later in this and subsequent chapters, are strictly only applicable to random samples, in that they are based on the assumption that individual observations have equal probabilities of being selected. On the other hand, the bias in many systematic samples is small, and may be a price worth paying for the time saved.

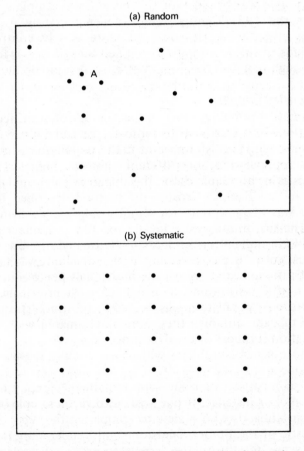

Figure 5.2. (a) and (b) Random and systematic point samples

A compromise between random and systematic methods is sometimes used, to simplify selection at a minimum cost in terms of bias. A good example is the procedure used in the 1966 British Sample Census of Population, in which 10 per cent of all households were enumerated. Each enumeration district was allocated a randomly selected digit, which we shall call n. The enumerator was then instructed to call only at the nth, $(10 + n)$th, $(20 + n)$th, etc. households as he walked systematically round his district. He was therefore spared the task of selecting his own sample randomly, while the bias of complete systematic sampling was minimized by varying the digit from district to district randomly. (If the digits had been 1 or 2 for every district, corner houses, which are often larger than their neighbours, would have been over-represented.) In fact, there is evidence that some enumerators did not carry out their instructions fully, so that the sample was biased. For this reason, the sampling was transferred, in the 1971 Census, from the enumerating to the data-processing stage, where it could be more closely controlled: for certain kinds of information, only 10 per cent of census forms were used, although they had been completed by all householders.

5.4c. Stratified sampling

A refinement of sampling procedure is the use of stratified samples, in which the population is divided into sub-sets, and separate samples drawn from each. This is important if the population is made up of distinct sub-populations of different size or character. Different procedures may be required for the different sub-populations. For instance, in the Sample Census, the method described in the last paragraph was inappropriate for enumerating people living in institutions. The institutions were locally too few in number and often too variable in number of residents to be treated in the same way as households, so a 10 per cent sample of individuals was taken from every institution, instead of all the individuals being counted in 10 per cent of the institutions, as happened with private households. The institutional population was treated as a distinct 'stratum' in the collection of data, although they were amalgamated with the rest of the population for most of the data processing.

Two more general problems which may require the use of stratified sampling are:

(i) The comparison of some aspect of the geography of two (or more) areas. For instance, if one wished to visit a sample of farms in the English Midlands with a view to comparing the types of farming characteristic of the Bunter Sandstone and Keuper Marl, it would be common sense to stratify the sample to ensure the selection of a roughly equal number of farms on each formation—in effect, one would take twoseparate samples. Otherwise there would be too few from one sub-area, or too many from the other.

(ii) The comparison of two variables, when their variability is not distributed in equal proportions over the area being studied. For instance, if one were investigating the relationship between altitude and vegetation on the limestone areas of the Peak District, it would be desirable to have adequate representation in each altitude band. But a simple random sample of points would be unlikely to provide this, because there is much more of the Peak District in some altitude bands (e.g. the thousand-foot surface) than in others. It could, however, be achieved by first stratifying the area into altitude bands of, say, 0–300 feet, 300–600 feet, 600–900 feet, etc., and then accepting the first 20 point in each altitude band that are selected by a random sampling procedure such as that described in section 5.4a. When a point is selected in a height band for which 20 points have already been found, it is passed over; the next random number is called, and the procedure continues as before until each height band has its full quota of 20 points.

Both examples in the previous paragraphs involve both stratification into sub-populations, and the random selection of sub-samples. They are therefore called 'stratified random samples'. If the selection of sub-samples is made systematically, then the result is called a 'stratified systematic sample'. The stratified sampling procedure adopted for the 1966 Census includes both random and systematic elements—yielding, one supposes, a stratified random systematic sample!

5.4d. Cluster sampling
The last formal sampling design to be described is cluster sampling. If the population is very widely dispersed, straightforward random or systematic sampling will involve a great deal of travel. Opinion polls in Britain, for example, generally use samples of about 2000 people, which is about one in every 25 000 of the population. It saves an immense amount of time to sample from carefully selected clusters— e.g. from ten inner urban areas, ten suburban areas, and ten rural areas of varying location, social structure, etc. A double sampling procedure is, therefore, involved—first, selecting representative clusters (which is probably best done subjectively) and then selecting a random, systematic, or stratified sample within each cluster.

A cluster technique can also be used to check on the reliability of a previously conducted survey. Again reference is made to the census: in this case to 1961, when a sample of 2500 small areas, each consisting of clusters of about 20 households and distributed all over the country, were revisited to check on coverage (was the number of dwellings, households, and persons the same?) and quality of information (were the answers on the census schedules the same?). The results were generally reassuring as to the reliability of the original enumeration.

We may now tabulate the various formal sampling designs that have been described as follows (Table 5.2). Many sampling procedures fit neatly into one of the cells in this table; others, such as those connected with recent censuses, are hybrids and belong to several cells.

TABLE 5.2
A classification of formal sampling designs

| | Spatial | | | |
	Point	Line	Area	Non-spatial
Random				
Systematic				
Stratified				
Cluster				

5.4e. Examples of sampling designs

Although sampling procedures are designed to minimize subjective selection, the choice of *sampling frame* (i.e. population or subpopulations from which the sample is selected) and *sampling method* (i.e. method of selection from the sampling frame) does require common sense. There is no set of rules that will cover every sampling problem and an investigator must be prepared to use his ingenuity in devising procedures that will provide reasonably unbiased data from a reasonable expenditure of time and effort.

To conclude this section, three examples will be given of sampling designs devised for particular problems:

1.   A student wished to obtain random samples of pebbles from different parts of a beach. So she adopted a procedure including both systematic and random elements. She walked along the cliff-top behind the beach, stopping at regular intervals. At each stopping point she looked down to her partner on the beach through a clinometer, which was set at an angle of declination determined by the use of random numbers (a different random number at each stop). Her partner was instructed to move up or down the beach to come

into line with her line of vision. When he was in line, he placed a wooden cross on the beach at his feet and picked up all the pebbles actually touching the cross.

2. Another student wished to investigate the relationship between distance from a city centre, rateable value, and function of building properties. To obtain his sample of buildings for scrutiny, he first took a series of three digits between 000 and 359 from the table of random numbers. These determined the bearings of each building in the sample from the city centre. He then took another series of three random digits from 0 to 180, to determine the distance of each sample building from the centre in tenths of an inch on a six-inch map of the city. (The furthest point of the built-up area from the city centre was 3 miles or 180 tenths of an inch on the map). Random bearings were then paired with random distances to give the locations of the buildings selected for the sample. Where the resulting location was in open country (outside the built-up area), it was ignored; where it landed in an open space within the built-up area, the nearest building to that location was chosen.

3. A random series of cross-sections were to be made across a relief map, to estimate the amount of land at different altitudes within the map area. To determine the line of each cross-section, two grid references were found, using random numbers, and a line drawn through the two points from one edge of the map to the other. The cross-sections were then drawn, and the lengths within each height band aggregated.

Consolidation
2. State the population and sampling frame in each of the three last examples, and any source of bias you can detect. Into which cell(s) in Table 5.2 would you place each?

5.5. THE USE OF SURROGATES
Sampling is used to ease problems of data collection. Another commonly used device to achieve similar ends is the use of surrogates i.e. substitute variables that tell approximately the same story as the variables we are really interested in (the target variables). We have already mentioned the counting of road intersections per unit area as a substitute for measuring the lengths of roads, when comparing the densities of road networks (p.63). Similarly, the number of contours cutting a unit length of grid line may be a surrogate for relative relief; wet-bulb temperature for potential evapotranspiration; school-leaving age or occupation for social class; unemployments rates for economic health, infant mortality for social well-being; newspaper circulation for literacy; distance for cost or time separating two locations.

In many ways surrogates are analogous to sampling:

(i) They facilitate data gathering, and in some circumstances are the only way of obtaining any data at all (when data on the target variable are not available, as in the case of illiteracy in many developing countries; or when the target variable is too diffuse to be pinned down to direct measurement, as with social well-being or economic health).

(ii) Surrogates are subject to random 'error' (i.e. variations in the surrogate will differ randomly from variations in the target variable) which may prevent an investigator achieving conclusive results, since the variations he is investigating are obscured by random variations in the surrogate. But whereas the level of random error in samples can be estimated from probability theory (as outlined in the next sections of this chapter), the same is not true of surrogates. Only if we have some target variable data with which to compare the surrogate data can we estimate the amount of random error in the latter; then, the higher the correlation between the two variables, the less random error there will be, and the more reliance can be placed on one variable as a surrogate for the other. For instance, travel cost correlates better with route distance than with straight-line distance. Therefore route distance is a better surrogate for cost than straight-line distance. (For correlation, see Chapter 7.)

(iii) Surrogates are also subject to bias (i.e. variations in the surrogate may differ systematically from variations in the target variable). Thus, use of occupations of heads of households to determine employment structure—as in the Census of Population between 1811 and 1831—will overestimate the kind of jobs undertaken by older men, and underestimate those undertaken by the young and by women. On the other hand, whereas in sampling there are standard procedures (as described in section 5.4) to minimize bias, the same is not true for the selection of surrogates—one simply has to rely on the application of common sense and geographical training to a knowledge of the factors relating a surrogate to its target variable.

One limitation of surrogates that involve measurement on an interval scale is that they generally yield data in different dimensions from those required for the target variable. Wet-bulb temperatures are not expressed in the same units as rates of evapotranspiration. (The limitation does not apply when target variable and surrogate are on nominal scales since the end product for both will be a frequency or percentage.) Therefore interval-scale surrogates can, as a rule, only be used to provide comparative and not absolute data on the target variable. This is often all that is needed, since much geographical study involves comparisons. If absolute data are required, the mathematical relationship between the surrogate and its target variable must be known—generally in the form of a regression equation (Chapter 8). An interesting example of this, which must be one of the earliest forays by a geographer into statistical techniques, and which predates the 'Quantitative Revolution' by more than twenty-five years, is Thornthwaite's use of regression analysis to obtain a workable classification of climates that used the ratio between precipitation and evaporation as one of its criteria (Thornthwaite 1933). While precipitation and dry-bulb temperature figures were widely available throughout the world, evaporation data were not. On the other hand, figures were available for precipitation, temperature, and evaporation over widely differing climates in the U.S.A., from Alaska to Florida. Thornthwaite used these American data to derive a regression equation expressing the precipitation-evaporation ratio (P.E.) in terms of precipitation (P) and temperature (T):

$$\text{P.E.} = 11 \cdot 5 \left(\frac{P}{T-10}\right)^{\frac{10}{9}}$$

(all figures in degrees Fahrenheit and inches).

He was then able to apply the P.E. criterion over the rest of the world.

In concluding these sections on sampling and surrogates, it should be emphasized that there is a pay-off between time and effort on the one hand, and precision on the other. Small samples and crude surrogates may save a lot of time but produce useless results. Very large samples and sophisticated measurements (whether of the target variable or of a surrogate) may produce unnecessarily high levels of accuracy, and therefore involve an unnecessary amount of time and effort. Large samples of the target variable are not always better than small samples of a surrogate. Neither is the reverse always the case. The investigator has to judge the optimum pay-off level for each problem.

5.6. POPULATION PARAMETERS AND SAMPLE STATISTICS
Samples are used to obtain estimates of numerical characteristics of populations, where it is impractical to measure or count the whole population. It is conventional to use the term 'population parameter' to denote a numerical characteristic of a population, and 'sample statistic' to denote a numerical characteristic of a sample. A sample

statistic (such as mean roundness of 100 randomly selected pebbles) generally provides the only estimate available of the corresponding population parameter (such as the mean roundness of all the pebbles on a beach). The most widely used parameters and statistics are the mean and standard deviation, together with the number of variates from which they are derived. They are indicated by the following conventional notation:

	Population parameter	Sample statistic
Number of variates:	N	n
Mean:	μ	$\bar{x}$
Standard deviation:	σ	s

5.7. SAMPLING DISTRIBUTIONS AND THE STANDARD ERROR

It is highly unlikely that any one sample, no matter how unbiased, will yield values of $\bar{x}$ and s which give precise estimates of the corresponding population parameters μ and σ. It is equally unlikely that two different samples, drawn from the same background population, will have exactly the same values of $\bar{x}$ and s. Indeed, if one took a large number of samples from the same population, it is probable that each would produce a different mean and standard deviation. A distribution of such sample statistics (e.g. of the means of samples drawn from a common background population) is known as 'a sampling distribution'. If the samples are all unbiased, it is reasonable to expect that their means will be distributed symmetrically about the corresponding population mean—some sample means will be bigger than the population mean, and a roughly equal number will be smaller. The same is likely to be true of other sample statistics in relation to their corresponding population parameters.

Now in practice we do not take a number of samples but only one, and we do not know the population parameters but can only estimate them from our one sample. Nevertheless, by *hypothesizing* sampling distributions based upon a large number of *imaginary* samples, mathematicians have been able to provide us with a measure of the accuracy of parameter estimates based on a single real sample. To understand how this is done, it is first necessary to introduce the central limit theorem and the concept of the standard error.

The central limit theorem states that if we imagine taking all possible samples of similar size from a single population, the sampling distribution of their means will be approximately normally distributed about the population mean, whatever the character of the population distribution, provided the samples are fairly large (say, over 30)—see Figure 5.3. Also, the standard deviation of the sampling distribution of means is given by: $\sigma/\sqrt{n}$, i.e. standard deviation of population/square root of size of sample.

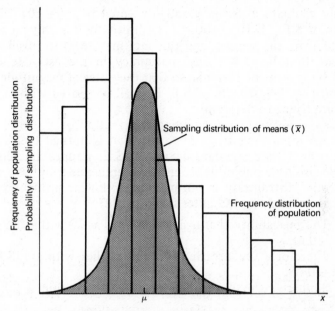

Figure 5.3. A sampling distribution of means derived from a skew population distribution

The standard deviation of a sampling distribution is called the standard error of the sample statistic. Therefore the standard deviation of the sampling distribution of means is called the standard error of the mean (S.E.$_{\bar{x}}$). Thus:

$$S.E._{\bar{x}} = \sigma/\sqrt{n} \hspace{3cm} \textit{Formula 5.1}$$

The standard error is the key to accurate estimation of population parameters from sample statistics, and how it is used for this purpose will now be explained.

5.8. ESTIMATES FROM SAMPLE MEASUREMENTS

We are now in a position to estimate likely *random error* in a sample statistic, by considering what it tells us about the corresponding population parameter. What follows takes no account of *bias*. Suppose we wish to estimate the mean roundness of pebbles on a stretch of beach. We take a random sample, and find that

$\bar{x} = 50$ and $s = 10$.

The value of $\bar{x}$ gives us a rough estimate of μ, but how rough? What is the likelihood of the true value of μ being as high as 60 or as low as 40?

We begin by estimating the standard error of the sample mean using Formula 5.1: $S.E._{\bar{x}} = 10/\sqrt{100}$. We know that $n = 100$, but can only estimate σ as being roughly equal to s (= 10). Thus our estimate of $S.E._{\bar{x}}$ is $10/\sqrt{100} = 1$. (Admittedly it is only an estimate obtained from the sample, and this fact may seem to invalidate the argument to follow. But any inaccuracy in the estimate of σ is divided by $\sqrt{n}$ in the formula, so that the order of magnitude of any inaccuracy in the estimate of S.E. is small compared with the mean (μ) we are trying to determine.)

We can now imagine a sampling distribution of $\bar{x}$, centred on μ (which we are trying to estimate), which is normal (central limit theorem) and has a standard deviation ($S.E._{\bar{x}}$) approximately equal to 1. We also have one value from this sampling distribution ($\bar{x} = 50$). Because the distribution is normal, the following probability statements can be made (see Chapter Four, p.122):

(i) The probability that a single value of x lies within 1 S.E. of μ is 0·68.
(ii) Therefore, the probability that 50 lies within 1 S.E. of μ = 0·68.
(iii) Similarly, the probability that 50 lies within 2 S.E.s of μ = 0·95.
(iv) And the probability that 50 lies within 3 S.E.s of μ = 0·997.

Now if A is a stone's throw from B, then B is a stone's throw from A. So the above statements can be reversed as follows:

(i) The probability that μ lies within 1 S.E. of 50 = 0·68.
(ii) The probability that μ lies within 2 S.E.s of 50 = 0·95.
(iii) The probability that μ lies within 3 S.E.s of 50 = 0·997.

And as $S.E._{\bar{x}} = 1$, we can restate this as:

(i) There is a 68 per cent probability that μ lies between $50 - 1$ and $50 + 1$, i.e. between 49 and 51.
(ii) There is a 95 per cent probability that μ lies between 48 and 52.
(iii) There is a 99·7 per cent probability that μ lies between 47 and 53.

These levels of probability are usually called 'confidence levels': and the limits within which a parameter lies with a specified level of confidence are known as 'confidence limits'. The range between confidence limits is called the 'confidence interval'.

Therefore, the 95 per cent confidence limits for pebble roundness on our beach are 48 and 52, and the 95 per cent confidence interval is 4. We can say with 95 per cent confidence that the mean roundness for the whole population of beach pebbles lies between 48 and 52. A value outside this range is sufficiently improbable to be ignored for many purposes, and the 95 per cent level of confidence is frequently regarded as giving tolerable limits for making population estimates.

More generally, we can state that:

The 68 per cent confidence limits of μ are $\bar{x} \pm 1$ S.E.
   95 per cent confidence limits of μ are $\bar{x} \pm 2$ S.E.
   99·7 per cent confidence limits of μ are $\bar{x} \pm 3$ S.E.

We may, however, wish to ascertain limits for confidence levels other than 68 per cent (which is too low for most purposes), 95 per cent (the most commonly used confidence level), or 99·7 per cent. We must then use the table of z-value probabilities associated with the normal distribution and introduced in the previous chapter (p.124). Confidence levels express the probability of values (from a sampling distribution which is normal) lying within specified limits on *either side* of the mean. The table of z-value probabilities in Appendix 1 col. A gives the probability that values taken from a normal distribution will lie within z standard deviations on *one side* of the mean. Therefore the probability associated with a z-value in the table is half the corresponding confidence level. The procedure for using z-value tables to determine confidence limits is therefore as follows:

  (i) Halve the required confidence level and express it as a probability out of 1 (e.g. for a 90 per cent confidence level, the probability required is 0·45).

  (ii) Look up the corresponding z-value in the table (for $p = 0·45$, $z = 1·6$).

  (iii) The confidence limits are then $\bar{x} \pm z$.S.E. In the case of the beach pebbles, where S.E. was about 1 and $\bar{x}$ was 50, the 90 per cent confidence limits for mean roundness are $50 \pm 1·6 \times 1$ i.e. 48·4 and 51·6.

At this point, it must again be stressed that the above methods of determining confidence levels are only valid if the sample from which the estimates are made is fairly large (over about 30); and that they take no account of possible bias.

Two further points about standard errors and confidence limits, which follow from what has been said:

(i) For a given size of sample, the standard error, and therefore the confidence interval, increases as the standard deviation of the sample increases—i.e. *the greater the standard deviation of the sample, the less accurately can one estimate the population mean.* This can be illustrated by considering how the 95 per cent confidence limits for mean roundness of our beach pebbles would vary with different values for a sample standard deviation (Table 5.3):

TABLE 5.3
The relationship between standard deviation, standard error and confidence interval

n	$\bar{x}$	s	$S.E. = \dfrac{s}{\sqrt{n}}$	*95 per cent confidence limits*	*95 per cent confidence interval*
100	50	10	1	48 and 52	4
100	50	20	2	46 and 54	8
100	50	30	3	44 and 56	12
100	50	40	4	42 and 58	16

The third and fifth columns of this table can be graphed as follows:

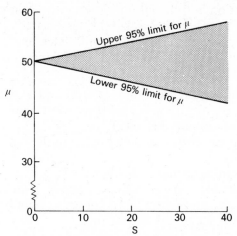

Figure 5.4. Diagram to show relationship between sample standard deviation and 95% confidence limits for μ

(ii) For a given standard deviation of sample, the standard error and confidence interval decrease, and the confidence limits narrow, as the sample size increases—i.e. *the greater the sample size, the more accurately one can estimate the population mean.* This can be illustrated by a table and graph analogous to those presented above.

Consolidation
3. Complete Table 5.4, and draw a graph of Columns 2 and 5, similar to that in Figure 5.4.

TABLE 5.4
The relationship between sample-size, standard error, and confidence interval (to be completed)

$\bar{x}$	n	s	S.E.	95 per cent confidence limits	95 per cent confidence interval
50	100	20	2	46 and 54	8
50	200	20			
50	300	20			
50	400	20			

4. In a survey among Indian villages, the diet of 100 randomly selected adult males was calculated to contain a mean daily calorific value of 2000, with a standard deviation of 250.
(a) Calculate the 95 per cent confidence limits for the mean daily calorific value of the diets of all adult males living in those villages.
(b) Assuming that s remains 250, by how much would the sample have to be increased in order to halve the confidence interval?

Estimates from small sample measurements
It will be recalled that the central limit theorem only applies to fairly large samples (over about 30). The larger the sample, the less the shape of a sampling distribution depends upon the normality of the background population. Thus, while the sampling distribution of large-sample means is normal, and this fact is used to estimate the population mean whatever the shape of the background population, the shape of the sampling distribution of small sample means depends upon the shape of the background population. If this is not known, there is no way of making a statistical estimate of the population mean from a small sample.

If we wish to estimate the mean of a *normally distributed* population from a small sample, two modifications to the procedure outlined above are necessary:

(i) It is found that a better estimate of the standard deviation of the population (σ) is obtained by multiplying the sample standard deviation (s) by $\sqrt{\{n/(n-1)\}}$. This is called Bessel's correction, and gives the 'best estimate' of σ, denoted by $\hat{\sigma}$. Thus,

$$\sigma \triangleq \hat{\sigma} = s \sqrt{\left(\frac{n}{n-1}\right)} \qquad \qquad \textit{Formula 5.2a}$$

or $\quad \sigma \triangleq \hat{\sigma} = \sqrt{\left\{\frac{\Sigma(x-\bar{x})^2}{n}\right\}} \sqrt{\frac{n}{n-1}} = \sqrt{\left\{\frac{\Sigma(x-\bar{x})^2}{n-1}\right\}}$

$$\textit{Formula 5.2b}$$

(It will be clear that, as n increases, $\sqrt{n/(n-1)}$ approaches 1, and the difference between s and $\hat{\sigma}$ becomes slight. That is why the modification can be ignored when n is greater than 30.)

(ii) The z-table used in making estimates from large samples needs to be replaced by what is called a t-table (Appendix 3) when the sample is small. This is because the sampling distribution of small-sample means is not normal, even if the parent population is normal, but follows what is called a 'Student's' t-distribution. ('Student' is the pseudonym for the statistician, W. S. Gossett, who discovered the t-distribution.) The shape of the t-distribution depends on the size of the sample, and as the latter increases, the t-distribution becomes more and more like a normal distribution.

The procedure for finding the confidence limits for the mean of a *normally distributed* population from a small sample will now be shown by an example.

A static count of pedestrians outside Woolworths on nine Saturday mornings gave a mean flow of 2500 per hour, with a standard deviation of 400. What are the 95 per cent confidence limits for the mean Saturday morning pedestrian flow?

Step 1: Find the best estimate of the standard deviation of the population from Formula 5.2:

$$\hat{\sigma} = s \sqrt{\left(\frac{n}{n-1}\right)} = 400\sqrt{(9/8)} = 424.$$

Step 2: Find the standard error of the sampling distribution from Formula 5.1:

$$\text{S.E.}_{\bar{x}} = \sigma/\sqrt{n} \triangleq \hat{\sigma}/\sqrt{n} = 424/\sqrt{9} = 141.$$

Step 3: Look up in the t-tables (Appendix 3) how many standard errors either side of the mean contain 95 per cent of the t-distribution, when the size of the sample is 9. This is done by reading off the figure in the 95 per cent column opposite *one less* than the sample size in the 'degrees of freedom' column. The figure in the table under 95 per cent and opposite 8 degrees of freedom is 2·306.

Step 4:   Multiply the standard error by 2·306

$$141 \times 2 \cdot 306 = 325$$

Step 5:   We can now give the 95 per cent confidence limits as 2500 ± 325;

i.e. there is a 95 per cent probability that the mean Saturday morning pedestrian flow outside Wooldworths lies between 2175 and 2825 per hour. It is worth noting that if the large sample method previously outlined had been used in this case, the confidence limits would have been calculated as 2234 and 2766—i.e. it would have given a false notion of the accuracy of the estimate.

Consolidation
5.   A number of farms in East Anglia were sampled to find wheat yields. The mean of the sample was 40 bushels per acre, and the standard deviation 5. Calculate by both the small sample and the large sample methods the 95 per cent confidence limits, (a) when the sample size was 16; (b) when the sample size was 25. Which method gives the correct confidence limits? What is the percentage error in the result of the incorrect method in each case?

5.9. ESTIMATES FROM SAMPLE COUNTS
So far we have only considered making estimates from samples in which each observation is a measurement—i.e. a number, such as pebble roundness, field acreage, or wheat yield, recorded on an interval scale. Often, however, the individual observations are counts rather than measurements. For instance, one may wish to determine the proportion of flint pebbles at each end of a beach. In this case, each individual observation is recorded on a nominal scale. Random samples are taken, in which each observation is recorded as either 'flint' or 'not flint'. These can then be counted up, and the proportion of flints in each sample calculated. But one cannot calculate a standard deviation for a single sample of counts, and cannot therefore derive the standard error and confidence limits by the methods previously outlined. One requires a different technique for making population estimates, using the standard error of the binomial distribution.

5.9a. The standard error of the binomial distribution

It will be recalled that the binomial distribution is associated with the repetition of events where there are only two possible outcomes —flint or not flint, arable or not arable, immigrant or native, and so on (see p.115).

If 40 per cent of the pebbles at one end of a beach are flint, then the probability of picking up a flint pebble at random is 0·4, and the probability of picking up a non-flint is 0·6. A random sample of 200 pebbles from this part of the beach should yield *about* 200 X 0·4 = 80 flints, and about 120 non-flints. More generally, if the probabilities of picking up flints and non-flints are p and q respectively (so that $p + q = 1$), then a random sample of size n should yield about np flints and nq non-flints. It is, however, unlikely that two samples from the same part of the beach would yield exactly the same results, and it can be shown that a large number of samples from the same population of pebbles would yield a sampling distribution for the frequency of flint pebbles, which would be approximately normal, with a mean of np, and a standard deviation (called again a standard error) of $\sqrt{(npq)}$, i.e.

$$\text{S.E.}_{np} = \sqrt{(npq)} \qquad\qquad \textit{Formula 5.3}$$

provided the sample size (n) is reasonably large. 'Reasonably large' depends upon the values of p and q. If $p = q = 0·5$, the binomial distribution resembles a normal distribution when n is quite small. The greater the difference between p and q, the greater n must be for the distribution to approach normal. A *rule-of-thumb is that* npq *should equal at least 9.* Therefore, if $p = q = 0·5$, n needs to be at least 36 (36 X 0·5 X 0·5 = 9); if $p = 0·1$ and $q = 0·9$ (or vice versa), n needs to be at least 100 (100 X 0·1 X 0·9 = 9).

The procedure for determining confidence limits from sample counts is therefore as follows:

  (i) Ensure that npq is at least 9, so that the sampling distribution is approximately normal. (If it is not, confidence limits cannot be estimated.)

  (ii) Calculate the mean (np) and standard error ($\sqrt{(npq)}$) of the sampling distribution.

  (iii) Use the z-table (Appendix 1 col. A) to find confidence limits in the same way as for large sample measurements (see p.145).

Applied to our 200 pebbles:

  (i) $npq = 200$ X $0·4$ X $0·6 = 48$, so that the sampling distribution is close to normal.

  (ii) Mean = 200 X 0·4 = 80; S.E. = $\sqrt{48} \approx 7$.

(iii) Confidence limits are $80 \pm z.7$

  e.g. 95 per cent confidence limits are $80 \pm 2 \times 7 = 66$ and 94

  99 per cent confidence limits are $80 \pm 2\cdot58 \times 7 = 62$ and 98.

5.9b. The binomial standard error as a proportion

However, when making population estimates from counts, one is generally interested in proportions rather than in actual numbers— one does not want to know how many flint pebbles there are on the beach, but what the proportion of flint pebbles is. It is therefore necessary to express the mean and standard error of the sampling distribution as percentages of the sample size (n). This is done by multiplying the numerical values by $100/n$.

Thus, the *mean* (np) becomes $np \times (100/n) = 100p$ per cent of n, and the *standard error* $(\sqrt{npq})$ becomes $\sqrt{(npq)} \times (100/n)$

$$= \sqrt{\left(npq \times \frac{100^2}{n^2}\right)} = \sqrt{\frac{100p \times 100q}{n}}$$

This is generally written as $\sqrt{(p\%q\%/n)}$, where $p\%$ and $q\%$ express the proportions p and q as percentages (e.g. 0·4 and 0·6 become 40 per cent and 60 per cent). Thus,

$$\text{S.E.}_{\cdot p} = \sqrt{\frac{p\%q\%}{n}} \text{ per cent of } n. \qquad \textit{Formula 5.4}$$

In the case of the 200 pebbles, the mean for flints becomes $100 \times 0\cdot4 = 40$ per cent of the sample; and the standard error becomes

$$\sqrt{(40 \times 60/200)} = \sqrt{12} \backsimeq 3\cdot5 \text{ per cent of } n$$

The confidence limits are therefore $40 \pm z.3\cdot5$, so that the 95 per cent confidence limits are $40 \pm 2 \times 3\cdot5 = 33$ per cent and 47 per cent. We can say with 95 per cent confidence that the proportion of flint pebbles on the part of the beach from which the sample was taken lies between 33 per cent and 47 per cent.

(The reader may have noticed an apparent fallacy in the above reasoning. We began by assuming a population value for p $(= 0\cdot4)$, the probability of picking a flint at random. In reality, we do not know the population value of p, but have to estimate it from the sample. It is unlikely that the sample will give us an estimate which is exactly correct. However, if the population value of p were 0·4, then there would be a 95 per cent probability that we should have estimated a figure between 0·33 and 0·47 from our sample of 200. Now suppose our estimate had been at one extreme of this range, so that we made $p = 0.47$. Then the standard error would have been $\sqrt{(47 \times 53/200)} = \sqrt{12.45} \backsimeq 3.5$. (Notice that there is little difference in the standard error.) We should then estimate the 95 per cent confidence limits as 47 ± 7 (i.e. as 40 and 54), meaning that, from our rather extreme example the true population value of p (0.4 or 40 per cent) lies right on the margin of our 95 per cent confidence limit range. Nineteen times out of twenty, our sample would give an estimate of p which is less extreme, and would therefore include the true value (0.4) within the 95 per cent confidence limits. The twentieth sample would fail to include the true value within the 95 per cent confidence limits estimated from it, which is why they are only 95 per cent and not 100 per cent confidence limits. The fallacy is more apparent than real.)

TABLE 5.5
The relationship between sample size, the standard error, and confidence interval

n	p	q	$100p$	$S.E. = \sqrt{\dfrac{p\% q\%}{n}}$	95 per cent confidence limits	95 per cent confidence interval
50	0.2	0.8	20%	$\sqrt{32} = 5.7\%$	*	*
100	0.2	0.8	20%	$\sqrt{16} = 4\%$	$20 \pm 8\%$	16%
200	0.2	0.8	20%	$\sqrt{8} = 2.8\%$	$20 \pm 5.6\%$	11.2%
400	0.2	0.8	20%	$\sqrt{4} = 2\%$	$20 \pm 4\%$	8%

* Where $n = 50$, npq is less than 9, so that confidence limits cannot be calculated.

Let us now consider how confidence limits and the confidence interval vary with the size of sample counts. A randomly selected group of people from a town contains 20 per cent who have lived there all their lives (i.e. are natives). Table 5.5 and Figure 5.5 show the 95 per cent confidence limits for the proportion of natives for different sizes of sample.

Consolidation
6. (a) Draw up a table and graph, to show the confidence limits of the proportion of natives in the town, if a sample of 100 contains (i) 80 natives; (ii) 60 natives; (iii) 50 natives; (iv) 40 natives; (v) 20 natives.
(b) From Figure 5.5 and from the graph you have drawn, state the relationship between confidence interval and (i) sample size, (ii) size of p.

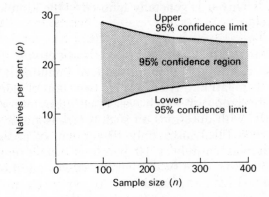

Figure 5.5. Graph to show relationship between sample size and 95% confidence limits of p from sample count

5.10. CORRECTION FOR SAMPLING FRACTION

The sampling fraction is the ratio between the size of a sample and the size of its parent population. If a sample of 100 is drawn from a population of 1000, the sampling fraction is $\frac{1}{10}$ or 10 per cent. Where the population is infinite or nearly so (as with point locations on a map), the sampling fraction is zero or near zero. In general, the precision of an estimate made from an unbiased sample depends on the latter's size rather than upon the sampling fraction, and the sampling fraction correction is· generally only made when the sampling fraction is more than one tenth.

If the sampling fraction is large, it reduces the standard error of the sample mean, and therefore of the confidence interval as well. The reduction is made by multiplying the standard error by $\sqrt{(1 - f)}$, where f is the sampling fraction. This applies, whether the standard error is derived from a sample of measurements, using $\sigma/\sqrt{n}$ (Formula 5.1), or from a sample of counts, using either $\sqrt{(npq)}$ (Formula 5.3), or $\sqrt{(p\%q\%/n)}$ (Formula 5.4). Thus, if the standard error is calculated by any of these formulae to be, say 5, and if the sampling fraction is $\frac{1}{4}$ (25 per cent), the corrected standard error is $5 \times \sqrt{(1 - \frac{1}{4})} = 5 \times \sqrt{0.75} = 5 \times 0.87 = 4.35$. This, is turn, reduces the 95 per cent confidence interval (for large samples) from 10 to 8·7.

The higher the sampling fraction, the greater the reduction of the standard error and confidence interval, until, when the sampling fraction is 1 (so that the whole population is sampled), the standard error and confidence interval are reduced to zero. On the other hand, if the sampling fraction is less than a tenth, the reducing factor is almost 1, i.e. $\sqrt{(1 - \frac{1}{10})} = 0.95$, and has little effect on the standard error. That is why it is generally ignored if the sampling fraction is less than a tenth.

5.11. THE STANDARD ERROR IN THE SAMPLE CENSUS

In 1961, the Registrar General introduced sampling into the British Census of Population. There were two householders' schedules (questionnaires), one of which was filled up by every householder, and the other, with questions on such topics as journey to work and occupation, was filled in by only 10 per cent of householders. The 1966 Census was entirely a 10 per cent sample enumeration. In 1971, all householders answered all the census questions, but many of the published data are based on a 10 per cent sample of the completed schedules. In all these cases the results are subject to sampling error.

The introduction to the published volumes of sample census tables states

For any sample total which is a small fraction (less than a quarter) of the whole sample population, the statistical quantity known as the standard error of this sample figure may be approximately estimated by the square root of the latter. To allow for the fact that sampling was on a 10% basis and was without replacement, this square root should be multiplied by the factor 0·9.

An example will clarify the meaning of this statement, the rationale for which will be found on p.354. In the 1961 Occupation, Industry, and Socio-economic Group tables for Nottinghamshire, the number of male textile workers *counted* in Nottingham County Borough is given as 903 (this is the 'sample total'), while the number of males counted in employment in all industries in Nottingham is given as 10,981 (this is the "whole sample population"). The sample total is therefore less than a quarter of the whole sample population, being about one twelfth. The standard error of the sample total may therefore be approximately estimated by:

(i) taking the square root of the sample total: $\sqrt{903} \simeq 30$
(ii) multiplying by 0·9: $30 \times 0.9 = 27$.

With a standard error of 27, we can now state the 95 per cent confidence limits of the sample total to be $903 \pm 2 \times 27 = 849$ and 957. As this was a 10 per cent sample, the true number of male textile workers in Nottingham in 1961 can be estimated with 95 per cent confidence to lie between 8490 and 9570.[1]

Consolidation
7. The 1961 and 1966 10 per cent Census tables give figures of

1,264 and 1,228 respectively for the numbers of female textile workers enumerated in Nottingham.

(a) Do the 95 per cent confidence limits for the true number of female textile workers in 1961 and 1966 overlap?

(b) What conclusion can be drawn from the answer to (a)?

5.12. ESTIMATING REQUIRED SAMPLE SIZE

Sampling involves a compromise between precision and economy of effort. A parameter can only be obtained with 100 per cent accuracy by measuring or counting the whole population. The smaller the sample, and, to a lesser extent, the smaller the sampling fraction, the wider the margin of error of any estimate made from it. (In fact, if the sampling fraction is less than one tenth, any variation makes little difference to the margin of error.)

If the margin of error that can be tolerated for a particular purpose is known, it is possible to calculate the minimum sample size required, thereby possibly saving time in the collection of an unnecessarily large sample. Different methods must be used, depending on whether the samples are to be measured or counted, but both involve the prior collection of a *pilot sample* of about 30 observations. These must be collected by the same methods that are to be employed in the full sample, and they can be used as part of the full sample, provided they do not affect the selection of additional items.

Sample size for measured data
The formula is:

$$n = \left(\frac{zs}{d}\right)^2 \qquad\qquad\qquad \textit{Formula 5.5}$$

where n is the required size of sample, s is the standard deviation of the pilot sample, d is the tolerable margin of error at a specified level of confidence, and z is taken from the z-table (Appendix 1 col. A) corresponding to the same level of confidence. (The margin of error is the interval between the mean and *one* confidence limit, and is therefore equal to half the confidence interval.) Therefore, if the pilot sample has a standard deviation of 10, and if the margin of error that can be tolerated is 2·5 at the 95 per cent level, then the required sample size is: $n = (2 \times 10/2\cdot5)^2 = 64$ (using Formula 5.5). Therefore, if the pilot sample contained 30 observations, an additional 34 would be needed to obtain an estimate within the margin of error that has been stipulated.

Sample size for counted data

This method is used where it is required to estimate the proportion of a population belonging to a specified category—e.g. the proportion of electors who will vote Labour. The formula is:

$$n = p\% \cdot q\% \cdot (z/d)^2 \text{ per cent}$$

where n, z, and d have the same meanings as before, and where p and q are the percentages of the pilot sample belonging and not belonging to the specified category respectively. Therefore, if 12 out of a pilot sample of 30 said they would vote Labour, and if one wished to estimate the proportion of Labour votes in the whole electorate with a margin of error of 1 per cent either way at the 95 per cent level, then the required size of sample would be: $n = 40 \times 60 \times (2/1)^2 = 9600$. (It is worth noting that most political opinion polls depend on samples of only about 2000, so that their margin of error is over 2 per cent, before counting factors such as bias and last-minute changes of mind on the part of the electorate.)

Consolidation

8. A pilot sample of 30 householders on an estate showed that income per householder had a standard deviation of £300 p.a., while 21 of the householders were car owners. What size sample would be required to estimate: (i) the average income per household on the estate, with a £50 margin of error at the 95 per cent level; (ii) the proportion of car owners on the estate, with a 5 per cent margin of error at the same level?

Quotations for discussion:

1. 'By the very nature of his interests the geographer deals only infrequently with samples. More often he is concerned with the total set of observations . . .' (King (1969), p.60). Do you agree?

2. 'The price of reaping these benefits (i.e. those accruing to probability calculus) is that we conceptualize geographic phenomena, generally speaking, as repetitive recurrent and independent events . . .' (Harvey (1969), p.265). What kind of geographic phenomena can justifiably be conceptualized as 'repetitive, recurrent and independent'? Is there any price to pay in doing so?

6
HYPOTHESIS TESTING

6.1. GENERAL CONSIDERATIONS

MUCH of a geographer's task involves the making of comparisons, both qualitative and quantitative. We are here concerned with the latter: the statistical comparison of places, areas, and areal distributions. The reasons for this preoccupation lie in the nature of geography as essentially a spatial discipline. Furthermore, comparison often points the way to explanation, which is always an important goal.

Comparison implies the identification of differences, similarities, and associations, which result from two groups of factors:

(i) Those which operate consistently and from which predictions can be made (e.g. the difference between land use in Wales and East Anglia is partly a function of rainfall, which is sufficiently predictable to influence the farmers of the two areas to make, in aggregate, different land-use decisions.

(ii) Those which are 'irregular'—i.e. which operate unpredictably and are often characterized as 'chance' or 'random' factors (e.g. the eccentricities of individual farmers may be partially responsible for deviations from regional norms).[1] The type of statistical comparison to be discussed in this chapter seeks to distinguish between these two groups of factors.

One particular situation which illustrates the interplay of these two groups of factors is the sampling process. If *exactly* 10 per cent of a town's population are pensioners, a random sample of the population can be expected to include *about* 10 per cent pensioners, but it is unlikely to include *exactly* 10 per cent. On the other hand, it is very unlikely (though not impossible) that the sample would include 90 per cent pensioners. The consistent factor (that 10 per cent of the total population are pensioners) will ensure that *about* 10 per cent of the sample will be pensioners; the inconsistent 'chance' factor (the 'luck of the draw' in the selection of individuals for the sample) makes it unlikely that *exactly* 10 per cent will be pensioners.

Now let us reverse the argument. Suppose that random samples of 100 people are taken from two towns, A and B, and that there are 9 and 18 pensioners in the samples from towns A and B respectively. It is unlikely that the proportions of the populations of the two towns who are pensioners are *exactly* 9 per cent and 18 per cent; but it is likely that they are about 9 per cent and 18 per cent. The question therefore arises: how precise is 'about'? Can we say with any confidence that the former town has a smaller proportion of pensioners than the latter? Or is it likely that the difference in the samples is entirely due to chance—i.e. to 'sampling error'? (See p.144). There is a standard procedure for answering this type of question, which will be outlined below, but it is first necessary to introduce the idea of the null hypothesis, commonly referred to in statistics as H_0.

6.1a. The null hypothesis (H_0)

One of the definitions of 'hypothesis' given by the *Concise Oxford Dictionary* is 'starting-point for investigation'. A null hypothesis is the starting-point of most statistical tests; it is a negative type of proposition, formulated for the purpose of applying a statistical test to a problem under investigation, and generally in anticipation of being rejected as false. In the above examples, we wish to investigate whether there really is a difference in the proportion of pensioners in the two towns, or whether the apparent difference is due to sampling error. There are two possibilities: (i) that there *is* a difference in the proportion of pensioners in the two populations; and (ii) that there is no difference. The latter is the negative proposition, and is the null hypothesis. More generally, there are two kinds of problem, for which different forms of H_0 are appropriate:

(i) If one is investigating the difference (and, by extension, the similarity) between two or more sets of data, then H_0 is of the form 'there is no significant difference between the data sets', which amounts to regarding them as samples from a single background population, so that any differences between them are no more than might be expected from random variations. This chapter is concerned chiefly with this type of problem.

(ii) If one is investigating the association between variations in data sets (as is the case with problems of correlation, to be discussed in the next chapter), then H_0 is of the form 'there is no association between the variation in corresponding values of the data sets.'

6.1b. Hypothesis testing

We can now return to the problem of the two samples, and outline the general procedure for handling it and other problems concerned with the difference between data sets.

Step 1 Formulate H_0. In the case of towns *A* and *B*, it could be stated as 'There is no difference in the proportions of pensioners in the two populations.' The point is that H_0 is formulated to calculate the probability that chance alone might yield the given data.

Step 2 Formulate an alternative hypothesis, H_1. In this case, it could be *either* that 'There is a difference in the proportion of pensioners in the two towns' (without specifying which town has the greater proportion); *or* 'Town B has a greater proportion of pensioners than town A.' The difference between these alternative hypotheses is important, and will be taken up a little later (p.164), but it need not worry us at the moment. The point to stress here is that H_1 must be formulated so that, if H_0 is rejected as a result of the statistical test to be applied (step 4), it is logical to adopt H_1 as an alternative.

Step 3 Decide upon a *rejection level* (α)—otherwise called a significance level: if the probability of the difference in the data occurring purely by chance under H_0 is less than α, then H_0 is rejected.[2] α is often set at 0·05, or 5 per cent, meaning that if differences at least as great as those of the problem data would occur by chance under H_0 less than five times in a hundred, then H_0 is so unlikely to be true that it is rejected, and the difference between the sets of data is said to be *statistically significant* at the 0·05 level. Another commonly used rejection level is 0·01, meaning that there is one chance in a hundred that the data being investigated could have occurred under H_0; or, putting it another way, in the light of the data, there is one chance in a hundred of the null hypothesis being true. The rejection level is a measure of how strong the evidence must be before H_0 is rejected. This should be decided by the investigator before the evidence is examined—i.e. the rejection level should be set before the statistical test is applied, reflecting the risk of a wrong conclusion.

Step 4 Select and carry out an appropriate statistical test to determine the probability (p) that the problem data could have occurred by chance under H_0. Enough tests are described in this and the following chapters to meet the requirements of most undergraduate students.

Step 5 If p (calculated in Step 4) is less than α (determined in Step 3) then H_0 is rejected at the p-level of significance. If p is greater than α, then H_0 cannot be rejected. This does not mean that H_0 has been proved correct, but simply that the evidence is not sufficiently strong to reject it. A larger sample might provide additional evidence by which H_0 could be rejected.

The formalization of the procedure outlined above is a useful aid in precise and logical thinking, and should be followed whenever a hypothesis is being tested statistically. Some additional comments may be useful:

(i) It can be applied to comparisons of whole populations as well as of samples. If a full count of towns A and B had revealed a difference in the proportion of pensioners, one may still have wanted to know the likelihood of such a difference occurring by chance. This can be done by treating the two town populations as if they were samples of a single imaginary statistical population, and then following a similar procedure to that just outlined. It is not suggested that a difference between two real populations *is* the result of chance, but the probability of a difference of this magnitude occurring by chance can be a useful yardstick for measuring its remarkability. For instance, a difference of 10 per cent in the proportion of pensioners (or steel workers) in the populations of two large cities can be shown, by following steps 1 and 4 above, to be more remarkable than the same difference in the populations of two small hamlets. Probability under H_0 provides a common scale for measuring the degree of remarkability of both differences. A difference which could have occurred by chance on only one occasion in a hundred is more remarkable than one which could have occurred ten times in a hundred. It may be a reason for further investigation of the former rather than of the latter.

Such a measure on the probability scale may be a useful *descriptive* statistic, whether or not it is used inferentially. Values of χ^2 and Correlation Coefficients (to name but two frequently used statistics) can be translated on to a common scale to describe quantitatively something about the relationship between data sets—viz. the remarkability of that relationship—whether those values are statistically significant or not.

Conversely, however, the significance scale does not measure the amount of difference or correlation, but only the likelihood of it occurring by chance. Highly significant differences and correlations are not necessarily large differences or high correlations, because of the importance of sample size in determining significance levels. Other things being equal, large samples yield higher significance levels than small samples. In fact, if samples are large enough, almost any difference or correlation can be found to be significant, because *significance levels only indicate the likelihood of there being some difference or correlation (however trivial) in the background populations. They say nothing about the magnitude of that difference or correlation, only about its 'remarkability'.* Descriptive statistics (such as the difference between two means or a correlation coefficient) and significance levels are complementary: the former measure the magnitude of differences and correlations in a collection of data; the latter their remarkability.

(ii) Nearly all statistical tests to be described make use of mathematically derived sampling distributions (p.144). Thus, for example, if two random samples are drawn from the same population, the difference between their means will approximate to zero; while the difference between the means of a large number of pairs of samples will tend to be normally distributed about zero, if each of the samples is sufficiently large (say, over 30). Positive differences occur when the first sample mean is greater than the second; negative differences when the second is greater than the first. Further, if the standard deviation of this sampling distribution (called the standard error of the difference-between-means) is known, the probability of a particular difference-between-means occurring under H_0 (i.e. in samples drawn from the same populations) can be determined, using z-tables. This is illustrated in Figure 6.1, which depicts a sampling distribution of difference-between-means $(\bar{a} - \bar{b})$ with a standard deviation of 4. As with all normal distributions, 95 per cent of the values (approximately) lie within the range of two standard deviations of the mean—i.e. within −8 and +8, since the mean is zero. Therefore the probability of obtaining a difference, under H_0, between two sample means which is outside this range is 5 per cent or 0·05, and is represented by the shaded areas under both extremities or 'tails' of the curve, added together; z-tables give probabilities associated with values lying outside other ranges.

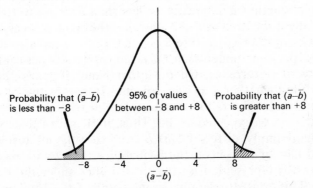

Figure 6.1. Sampling distribution of the difference-between-means $(\bar{a} - \bar{b})$ with a standard deviation of 4

(iii) It will be noticed that the shaded area in col. B of Appendix 1 is at one extremity of the normal curve, whereas in col. A the shaded area is adjacent to the mean. This is because col. A is used to calculate confidence limits, and for this purpose we need to know the probabilities associated with values lying *within* different distances of the mean of the sampling distribution; while col. B (and col. C) are used for testing hypotheses, when we want to know the probabilities associated with values lying *beyond* different distances of the mean of a sampling distribution. Therefore, when calculating confidence limits with z-tables, Appendix 1A is used, and probabilities (confidence levels) are usually of the order of 95 per cent or 99 per cent, and when hypotheses are tested with the aid of z-tables, Appendix 1B or C is used, and probabilities (significance levels) are of the order of 5 per cent or 1 per cent, more generally expressed as 0·05 or 0·01. Of course, not all hypothesis tests make use of z-tables, but the same principle applies with other sampling distributions.

(iv) When H_0 is of the type 'there is no significant difference between the data', the alternative hypothesis (H_1) can take one of two forms:
   (a) non-directional—in which only the fact of there being a difference between the populations represented by the data is stated.
   (b) directional—in which it is specified which population has the greater value of whatever the data represent.

Referring again to Figure 6.1, if H_1 is non-directional, it is not specifying whether $\bar{a} - \bar{b}$ is expected to be positive or negative. A difference between $\bar{a}$ and $\bar{b}$ can be due to $\bar{a}$ exceeding $\bar{b}$ (so that $\bar{a} - \bar{b}$ is positive), or to $\bar{b}$ exceeding $\bar{a}$ (so that $\bar{a} - \bar{b}$ is negative). The former case is represented by the right-hand half of the curve, the latter by the left. Thus, a difference of more than 8 which does not specify whether $\bar{a}$ is greater or less than $\bar{b}$, is, as we have seen, represented by the area under both tails (the shaded areas in Figure 6.1), which together constitute 5 per cent of the total area under the curve, (2·5 per cent under each tail), indicating a 5 per cent or 0·05 probability of occurrence. On the other hand, if H_1 is directional, it is specifying whether $\bar{a} - \bar{b}$ is expected to be positive or negative, and therefore also the tail of the sampling distribution in which the value of $\bar{a} - \bar{b}$ is expected to lie. Thus, if H_1 states that $\bar{a}$ *exceeds* (rather than simply differs from) $\bar{b}$, we are only interested in the right-hand half of the curve; the probability that $\bar{a}$ will exceed $\bar{b}$ by more than 8 is represented by the shaded area under the right-hand tail only. This constitutes only 2·5 per cent of the total area under the curve, denoting a probability of occurrence of 2·5 per cent or 0·025. Similarly, if H_1 states $\bar{b}$ to be greater than $\bar{a}$, then the probability of that difference being greater than 8 is represented by the shaded area under the left-hand tail of the curve (($\bar{a} - \bar{b}$) is less than, i.e. further from zero than, -8). This is also 2·5 per cent or 0·025. Therefore the probabilities associated with a directional alternative hypothesis are 'one-tailed' and only half the 'two-tailed' probabilities associated with a non-directional alternative hypothesis. It follows that, where H_1 is directional, a *one-tailed test* is called for; where H_1 is non-directional, a *two-tailed test* is required.

In our example of the proportion of pensioners in towns A and B, the percentages in the two samples were said to be 9 and 18 respectively. It can be shown that the standard deviation of the sampling distribution of differences (i.e. the standard error of the difference) is about 5 for samples of 100 from each town (for how this is calculated, see p.152). Therefore, the difference between the sample values is 9, which is 1·8 times the standard deviation of the sampling distribution—i.e. the difference can be expressed by a z-value of 1·8. If H_1 is 'Town B has a greater proportion of pensioners than town A', it is directional; a one-tailed test is therefore appropriate, and the one-tailed probability associated with $z = 1·8$ is 0·036. If, on the other hand, H_1 is simply 'There is a difference in the proportion of pensioners in the two towns', it is non-directional; a two-tailed test is required, and the two-tailed probability associated with $z = 1·8$ is 0·072 (i.e. twice 0·036). It follows that, if the rejection level has been set at 0·05, then H_0 would be rejected with a one-tailed test, but not by a two-tailed test.

Clearly, a great deal may depend upon the nature of the alternative hypothesis, since it determines whether a test should be one- or two-tailed, and this may determine whether or not a null hypothesis is to be rejected. An alternative hypothesis, requiring a one-tailed test, is statistically less demanding (i.e. H_0 is more easily rejected) and should only be formulated if there are strong reasons, *external to the data being tested*, to justify it. (e.g. if town A were Basildon and town B Bournemouth, there would be good *a priori* reasons for expecting town B to yield a sample with a higher proportion of pensioners, because New Towns generally have a smaller proportion of old people than seaside resorts. Therefore a directional H_1 would be justified.) If in doubt, it is wise to formulate a non-directional alternative hypothesis, and apply a two-tailed test. In the last resort, the choice between a directional and a non-directional alternative hypothesis is, like the selection of the rejection level (α), arbitrary; it is based on common sense and experience rather than on mathematical logic.

(v) There are two important thresholds which must not be confused when we are testing for significance. One is the rejection level (α). The other is called 'the critical value'. This is the value of the statistic being used to test some data (such as a z-score), beyond which the result of the test is significant at a stipulated rejection level. Thus, if z-tables are being used in a two-tailed test, and the rejection level is 0·05, then the critical value of z is 1·96, since any value of z which is *greater* (more extreme) than 1·96 has a probability of occurrence which is *less* than the rejection level of 0·05. The statistic being used sometimes has to be *greater* than the critical value for it to be significant (as with the z-score) and sometimes *less* (as with U and T in the Mann-Whitney and Wilcoxon tests)—it depends on which part of the sampling distribution has been used to calculate the table of critical values. The probability associated with a calculated statistic must always be *less* than the rejection level.

(vi) One of the chief values of probability statistics is that they narrow down the subjective element in an investigation. Measurement and calculation replace 'eyeballing' and guesswork. But subjectivity cannot be entirely eliminated. Its main role in hypothesis testing is in the determination of the rejection level and, to a lesser extent, whether to apply a one- or two-tailed test. We have noted that it is often set at either 0·05 or 0·01, but these are simply points on a continuum, and there is nothing intrinsically special about them. Like the pass mark in an examination, the precise location of the rejection level is arbitrary. But it is necessary if firm conclusions and practical decisions are to follow from an investigation whose results are not 100 per cent conclusive (as is generally the case in geographical investigations). If the rejection level is undemanding (say at 0·10) there is too great a danger of rejecting the null hypothesis when it is really true. This is called a 'type I error' by statisticians. Conversely, if the rejection level is too severe (say at 0·001) there is too great a danger of failing to reject the null hypothesis when in fact it is false. This is called a 'type II error'. Avoiding either a type I or a type II error is generally best assured by setting the rejection level at a happy medium of between 0·01 and 0·05.

(vii) *Independence.* Most probability theory assumes, among other things, that the data to which it is being applied are 'independent' of each other. All hypothesis tests stipulate this as a requirement, even those which, like correlation, involve paired (and therefore not independent) variables. The reason for this apparent contradiction is that the term 'independence' is used in a variety of ways by statisticians, and it is therefore necessary to distinguish between them. The several meanings may be conveniently explained by stating what kind of data are *not* independent.

(a) Autocorrelation: this refers to a purely arithmetical relationship between variates in one or more data sets. For instance, the *percentage* of population in one age group will, other things being equal, directly affect the percentage in other age groups (whereas the *numbers* in the age groups are not related in this purely arithmetical way). Or again, the length of a slope directly affects its height—the latter is a simple function of the former, other things being equal (whereas there is no such relationship between length and angle of slope). One meaning of independence is *freedom from autocorrelation*. The number (but not the percentage) in two or more age groups may be said to be independent; so may the length and angle of slope (but not length and height of slope). It is in this sense that the term is generally used when it is stated that a hypothesis test is only valid for independent data.

(b) Related events: when the probability of A affects the probability of B, they cannot be regarded as independent. For instance, the occurrence of drought in one year may affect the probability of its occurrence in the following year. If it can be shown, either statistically or by an analysis of meteorological processes, that probabilities of individual droughts are related, then they are not independent. Events are said to be independent if the chances of their occurring are not affected by each other. The probability Laws of Addition and Multiplication (p.112) can only be applied to independent events.

(c) Correlated data: these are similar to related events. A correlation coefficient is a measure of the extent to which a score on one variable (one event) is related to (statistically, not necessarily causally) a corresponding score on another variable (another event)—see Chapter 7. Significantly correlated data refer to related events; the lower and less significant the correlation, the greater the likelihood that the data, the events to which they refer, and the variables they represent, are independent.

(d) Matched data: when comparing data sets, it is sometimes possible to match individual scores from each set; when correlating data sets, it is essential to do this. For example, unemployment rates for a group of towns at two different times yield two figures for each town. Such data sets are said to be matched or paired (see p.188 for further discussion). Unmatched data are often referred to as independent.

6.1c. Criteria for the selection of statistical tests
The choice of tests for comparing the difference between sets of data (associations are dealt with in the next chapter) depends on three factors: (i) characteristics of the data, (ii) the relative 'cost-effectiveness' of different tests, (iii) assumptions that can be made about the background populations from which the data are derived.

6.1ci. Characteristics of the data   Characteristics of the data limit the choice of tests than can be applied to a problem. Among these, the following are especially important: (i) They may arise from either *counts* of items with different qualities or from *measurements* of different magnitudes. (ii) Measurements may be expressed on either an *ordinal* or an *interval* scale (see p.xvi). Observations differentiated on a nominal scale are regarded as counts rather than as measurements. (Strictly and statistically speaking, 'counts' are also measurements as explained on p.xvii. Each count is a measurement of one unit. But it is useful to make the above distinction, which is no doubt why we do so in everyday English.) (iii) Data may be paired (each count or measurement in one data set is matched with a corresponding count or measurement in the others) or unpaired.[3] In addition, one may be comparing a single real data set (or numerical distribution) with a hypothetical distribution, or two real data sets; or more than two data sets.

Some indication of the limits on choice of test is given in Figure 6.2, which shows the situations in which the tests described in the latter half of this chapter can be applied. It is clear that in some situations there is a very restricted choice of tests available, although it must be remembered that there are many more statistical tests not described in this book than are described. It will be observed that there is greater choice when data are on an interval scale, because such data can be converted into ordinal or nominal data, whereas the reverse is not possible.

6.1cii. The cost-effectiveness of different tests   The 'cost' in this context means the necessary effort in data collection and calculation. 'Effectiveness' means what statisticians call 'power'—i.e. the ability to discriminate correctly between a true and false hypothesis. Power depends in part on the size of samples being tested, and in part on the 'power efficiency' of the test. A test of low power-efficiency requires larger samples to achieve the same level of discrimination (i.e. power) as a test of high power-efficiency. For any given statistical problem there will be one test which is more power-efficient than any other that can be applied to it—i.e. it will require smaller samples to achieve a given power than any other test. If the sample size required to achieve a given power level by the most powerful (i.e. power-efficient) test applicable to that situation is N_1, and if the sample size required by another test to achieve the same power level is N_2, then the *power-efficiency* of the latter test $= N_1/N_2 \times 100$ per cent. Therefore, a test with a power-efficiency of 80 per cent (four-fifths) will require a sample size 125 per cent (five-fourths) of that required by the most powerful test available.

	Raw data		Transformed into (where shown)	Test
Comparison of two sets of data	Counted to give frequencies on nominal scale	In two categories	→	Binomial or chi square
		In more than two categories	→	Chi square
	Unpaired measurements (generally on an interval scale)		→ Frequencies →	Chi square
			→ Ranks →	Mann–whitney
			→	t – test
	Paired measurements (on an interval scale)		→ Ranks →	Wilcoxon
			→	t-test
Comparison of more than two sets of data	Counted to give frequencies on a nominal scale		→	Chi square
	Unmatched measurements (generally on an interval scale)		→ Frequencies →	Chi square
			→ Ranks →	Analysis of variance (Kruskal–wallis)

Figure 6.2. Tests to compare the difference between data: suitability related to characteristics of the data

Generally speaking, high-powered tests require more carefully measured data, lengthier calculations (unless a computer is available), and make more assumptions about the background population (see next paragraph) than do low-powered tests. In choosing between two tests therefore, one has to weigh these disadvantages of the more powerful test against the fact that the less powerful test will require the collection of larger samples. It is useful, in weighing these factors, to know the power efficiency of available tests, and this is given for each of the tests described hereafter.

6.1ciii. Assumptions about the background population   Statistical tests may be divided into two families: (i) 'classical' or 'parametric' tests, which dominated statistical theory and practice until recently, and which can only be applied to data on an interval scale: and (ii) "distribution-free' or 'non-parametric' tests, which have become important since the 1939–45 War, and which can be applied to data on nominal and ordinal as well as interval scales. Classical tests are generally more powerful, but they make certain assumptions about the background populations from which samples are drawn, which may not always be warranted, and which certainly should not be taken for granted. The most frequent of these is that the background population is approximately normally distributed, and the smaller the samples being tested, the more nearly normal must the background population be for most parametric tests to be valid. If therefore the normality or near normality of the background population cannot be reasonably assumed, then a 'distribution-free' test (which makes no assumptions about the distribution of the background population) should be used, and this is especially important if the samples being tested are small. Of the tests described in this chapter, only the *t*-test is classical; the rest are distribution-free.

6.2. THE χ^2 (CHI SQUARE) TEST

The χ^2 test, and the binomial test which follows, are methods for comparing *counted* data, in which individual observations are assigned to categories (i.e. are differentiated on a nominal scale), and the number in each category counted. Both tests are essentially different in this respect from others in this chapter, in which individual observations are recorded as *numbers*, on either an interval or an ordinal scale.

χ^2 is a simple technique which works by testing a distribution actually observed in the field against some other distribution determined by H_0—for example, that the objects under study are evenly spread over the landscape.

Requirements for the use of the χ^2 test are:

(a) The data must be in the form of *frequencies* counted in each of a number of categories (percentages cannot be used).
(b) The *total* numbers observed must exceed 20.
(c) The *expected* frequency under H_0 in any *one* fraction must not normally be less than 5. (See also p.174 Note 1.)
(d) The observations must be independent (i.e. one observation must not influence another).

It is normally not possible to give a meaningful estimate of the power efficiency of χ^2, because the test is generally used if alternatives are not available or are of doubtful validity.

The formula is simple:

$$\chi^2 = \sum \frac{(O-E)^2}{E} \quad \text{where } O = \text{the frequencies actually observed}$$

$$\text{and} \quad E = \text{the frequencies expected.}$$

Formula 6.1

χ^2 is, therefore, a measure of the aggregate difference between observed frequencies and those expected under H_0, so that the greater its value the less likely it is that H_0 is correct.

The application of the formula is best shown by means of an example.

6.2a. *Example 1*

The problem was to determine whether, in an area of south Devon, the distribution of farms relying mainly on milk production was influenced significantly by differences in surface geology.

Procedure

From the 1 inch to 1 mile geology map equal areas were selected consisting of the following rock types: Keuper Marl, Upper Sandstone, Upper Marl, Clay-with-Flints, Culm Measures. All farms within each area were visited. Any farm obtaining 75 per cent or more of income from milk was classified as 'mainly relying on milk production'.

Merely by inspection of the data shown in Table 6.1a it might well be inferred that rock type plays a major part in the decision of the farmer to produce milk. But before we can draw any conclusions, that is before we can begin to interpret the evidence, we have first to determine the probability that such a distribution could be due to 'chance'. The data are on a nominal scale (obtained by sub-dividing the farms into classes according to the type of rock on which they are situated), and are in the form of *frequencies*. The sample size is

TABLE 6.1a   *The χ^2 (Chi Square) Test*

Area	Rock type	Number of farms relying mainly on milk
1	Keuper Marl	8
2	Upper Sandstone	10
3	Upper Marl	6
4	Clay-with-Flints	3
5	Culm Measures	3
		30

small and the total distribution would almost certainly be markedly different from the normal curve. A parametric test is therefore not appropriate, but a test such as χ^2 can be used.

Information is also necessary about the *size* of the farms. A test would not be meaningful if, for example, the three farms on the Culm Measures were of a larger total area and with high production of milk than the ten farms on the Upper Sandstone. In fact this was an area of relatively small farms with no considerable variation in size.

1. Our null hypothesis, H_0, is that the frequency of farms depending mainly on milk for an income is not related to rock type (even though certain formations might be expected to carry a soil better suited to the production of grass than others).

2. H_1 states that there is a significant relationship between one rock type and another in the number of farms relying mainly on milk.

3. The rejection level for H_0 was decided at $\alpha = 0 \cdot 05$.

4. The numbers of farms in Table 6.1a represent the 'observed' frequencies: that is, the actual numbers counted in the field. If the local geology did not have any influence on the farmers' decision, it would be reasonable to suppose that we should find an approximately even distribution of farms irrespective of rock type. To obtain our 'expected' frequencies, therefore, we have to calculate the number of this kind of farm we should expect to find in each area if they were distributed evenly. We have in this case a total of 30 farms in 5 areas of equal size. Our 'expected frequency' in each area is therefore 6. We now use χ^2 to test the probability that the observed frequencies are so different from the expected frequencies that they are unlikely to be due to 'chance' variations. χ^2 is a measure of this difference. Obviously the greater the variation between what we 'expect' and what we 'observe', the larger will be the value of χ^2, and the greater will be the probability that we can reject H_0.

To find the value of χ^2 it is now possible to use the Formula 6.1

$$\chi^2 = \sum \frac{(O - E)^2}{E}$$

where O is the observed frequency and E is the expected frequency.

Every area over which we have conducted our survey now provides one fraction in the equation, thus:

TABLE 6.1b

Rock type	Observed frequency	Expected frequency	Formula
Keuper Marl	8	6	$(O - E)^2/E = (8 - 6)^2/6$
Upper Sandstone	10	6	$(O - E)^2/E = (10 - 6)^2/6$
Upper Marl	6	6	$(O - E)^2/E = (6 - 6)^2/6$
Clay-with-Flints	3	6	$(O - E)^2/E = (3 - 6)^2/6$
Culm Measures	3	6	$(O - E)^2/E = (3 - 6)^2/6$

χ^2 is therefore equal to:

$$\frac{(8 - 6)^2}{6} + \frac{(10 - 6)^2}{6} + \frac{(6 - 6)^2}{6} + \frac{(3 - 6)^2}{6} + \frac{(3 - 6)^2}{6}$$

$$= 0.67 \ldots + 2.67 \ldots + 0 + 1.50 \ldots + 1.50 \ldots$$

$$= 6.34 \ldots$$

(Note that any minus signs have been eliminated because when a negative number is squared the result is always positive.)

We now have a value for χ^2 of about 6.3 which can be referred to Appendix 4. It will be seen from the table that 'degrees of freedom' or 'df' have also to be calculated for each test. These are simply one less than the total number of fractions (individual values of $(O - E)^2/E$) involved, i.e., $df = n - 1$, where n is the number of fractions in the test. In this test we have five fractions. Therefore $df = 5 - 1 = 4$ (see p.180 for further explanation).

It is now possible to assess the extent to which the distribution we have been testing may be due to 'Chance'. Appendix 4 shows that with 4 degrees of freedom the critical value of chi square is 9.49 at the 0.05 level of significance. The calculated value of χ^2 in this case is 6.34. As this is less than the tabled chi square value at the 0.05 level it indicates that there is a more than 5 per cent probability that the distribution we have tested is random. Or, in other words, one might expect such a distribution to occur purely by chance more than five times in one hundred. We cannot therefore reject the null hypothesis. In spite of the initial impression our conclusion can only be a negative one, that on the basis of the data available there is insufficient evidence to assert that reliance on milk production as a main element of farm income is affected by rock type.

Note 1 If only two categories are involved (i.e. with one degree of freedom) the *expected* frequency in any fraction must not be less than 5. If more than two categories are involved (i.e. with $df > 2$) then not more than 20 per cent of the expected frequencies may be less than 5, and in no case must any expected frequency be less than 1.

Note 2 'χ^2' is used to indicate the value found by calculation from the observed data. The term 'chi square' is reserved for the critical values shown in Appendix 4.

Summary
1. Formulate H_0 and H_1. Decide upon the rejection level.
2. Tabulate the observed frequencies (O) and calculate the expected frequencies (E).
3. Using Formula 6.1 find the value of χ^2.
4. Obtain the significance of the result from Appendix 4.

6.2b. Example 2
In considering Example 1 the areas of land of each rock type for which the survey was made were equal, and the expected frequency was found by dividing the number of farms by the number of areas. The use of equal areas makes calculation simple, but is very often impracticable in the field, or unsuited to the problem in question. Such an example occurred in the course of an analysis of the drainage development in the basin of the upper Trent.

The drainage basins of the Trent and its tributaries upstream from Nottingham have developed over varied geological and morphological conditions. Visual inspection of the map (Figure 6.3) and calculated indexes of stream frequency indicate that these differences are reflected in the development of drainage. The problem is this. Are the differences between the basins so great that they are unlikely to be the result of chance? Unless this is established first, arguments concerning reasons for apparent differences are greatly weakened.

It was decided to analyse the development of drainage by identifying initial tributary streams in four main tributary basins and the headwater area of the River Trent (Figure 6.3). The data would be on a nominal scale of measurement (i.e. the streams would be categorized by drainage basins), and in the form of frequencies, so a chi square test was chosen as the most appropriate to determine whether

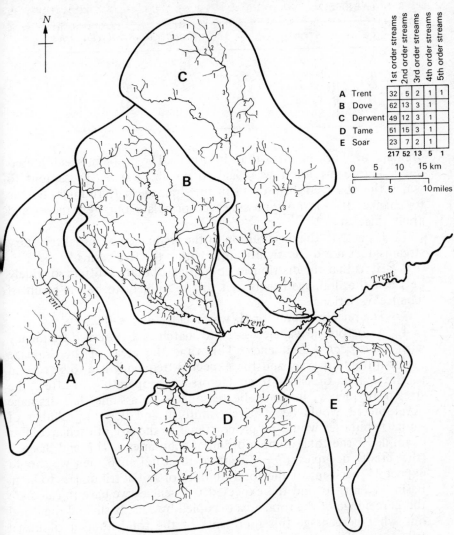

		1st order streams	2nd order streams	3rd order streams	4th order streams	5th order streams
A	Trent	32	5	2	1	1
B	Dove	62	13	3	1	
C	Derwent	49	12	3	1	
D	Tame	51	15	3	1	
E	Soar	23	7	2	1	
		217	52	13	5	1

Figure 6.3. Tributary streams in the five main basins of the upper Trent drainage area

variations in the numbers of these streams within the five basins could be regarded as statistically significant.

Table 6.2a gives the number of first-order streams counted in each basin. These were our observed frequencies. The five basins, not all of which had the same erosional history, also contain lithological and structural differences (we are not concerned with the precise nature of these here) which appeared from the map to have led to a denser

TABLE 6.2a

Area		Number of initial tributaries
A	R. Trent	32
B	R. Dove	62
C	R. Derwent	49
D	R. Tame	51
E	R. Soar	23

drainage network in some than in others. The problem was to determine, before attention was paid to possible causes, the extent to which the observed differences in the drainage pattern might be due to chance. It was now possible to formulate our hypotheses and apply this test.

1. H_0 is that 'the distribution of the frequency of initial tributary streams does not differ significantly within the five basins.'

2. H_1 is that the distribution of initial tributaries differs so widely between the drainage basins that this is unlikely to be the result of 'chance' variations.

3. The rejection level for H_0 was decided at $\alpha = 0.05$.

4. As before, it was necessary to match each *observed* frequency with an *expected* frequency. This time the areas of the drainage basins differed in size, and the expected frequency had to be related to the extent of each. If no factors were operating to affect the stream frequency, it would be reasonable to assume that drainage development would be relatively uniform and that the number of initial tributaries would be proportional to the size of each area. For example, if the total number of initial tributaries is 217, and Basin A (the Trent) occupies 17 per cent of the total area, then we should expect 17 per cent of the total number of initial tributaries to lie in Basin A. Thus to find the expected frequencies we have to calculate the percentage of the total area occupied by each basin and then find out what percentage this represents of the total number of initial tributaries. The result is set out in Table 6.2b. The value of x^2 is calculated as before.

$$x^2 = \sum \frac{(O-E)^2}{E}$$

$$= \frac{(32-36)^2}{36} + \frac{(62-50)^2}{50} + \frac{(49-65)^2}{65} + \frac{(51-39)^2}{39} + \frac{(23-27)^2}{27}$$

$$= 11.54 \ldots$$

TABLE 6.2b

Drainage basin	Units of area	Per cent of total area	Number of initial tributaries	Expected frequency
A	260	17	32	36 (i.e. 17% of 217)
B	355	23	62	50
C	466	30	49	65
D	286	18	51	39
E	193	12	23	27
	1590	100	217	217

5. There were five fractions in the calculation, giving four degrees of freedom ($df = 5 - 1 = 4$). Reference to Appendix 4 shows that with $df = 4$ a chi square value of 9·49 is significant at the 0·05 level. As the calculated value of χ^2 was 11·54, H_0 could be rejected, because the probability that the observed distribution might have occurred by 'chance' was less than 5 in 100. We are therefore able to accept H_1 and to state that the frequency of initial tributaries in the five drainage basins differs significantly at the 0·05 level of significance. Having established that apparent differences are significant, the way is clear for further research into the reasons why this should be so.

The individual values for each cell are also important, because they show whether certain cells (and therefore certain classes of data) contribute more than others to the value of χ^2, or whether this is the reflection of a *general* trend. In this case the basins of the Dove, Derwent, and Tame differ widely from the expected frequency, whilst those of the Trent and Soar roughly approximate to it.

Summary
1. Formulate H_0 and H_1. Decide upon the rejection level.
2. Tabulate the observed frequencies (O). Calculate the expected frequencies (E) by expressing each area under consideration as a percentage of the total area, and allotting E as that percentage of the total observed frequencies.
3. Using Formula 6.1 calculate the value of χ^2.
4. Obtain the significance of the result from Appendix 4.

6.2c. Example 3
So far we have considered the application of χ^2 to *one* variable (e.g. first farms, and then streams). The use of the test for two (or more) independent samples is also possible. The example chosen this time

also involves measurements on an ordinal scale, although for the purposes of this test they are treated as nominal.

The problem was to determine whether the angles of natural slopes (having the same aspect) which had developed on Keuper Marl and Greensand in two adjacent areas were significantly different. As the erosional history and vegetation cover of both areas was similar, it may be inferred that significant differences of slope angle result from the nature of the rock itself.

Measurements were taken randomly in the field using an Abney level with an attempted accuracy of one fifth of a degree. The results were then grouped in 5 degree intervals and tabulated as follows:

TABLE 6.3a

		Number of slope measurements	
		Keuper Marl 1	Greensand 2
1	0–4·8	21	9
2	5–9·8	18	11
3	10–14·8	9	15
4	15–19·8	8	13
5	20–24·8	1	5

1. H_0 is that no significant difference exists between the natural angles of slope which have developed upon the two kinds of rock.

2. H_1 is that the differences in slope angle are so great as to make it unlikely that they are the result of 'chance' variations.

3. The rejection level was decided as $\alpha = 0.05$.

4. Data were in the form of frequencies and χ^2 was chosen to test H_0.

The formula is very similar to that used previously:

$$\chi^2 = {}^r\Sigma^k\Sigma \ (O-E)^2/E \qquad\qquad \textit{Formula 6.2}$$

The usual symbols used for 'rows' and 'columns' are r and k. ${}^r\Sigma^k\Sigma$ means 'sum over all rows and columns' (i.e. sum all the cells in the table). The observed frequencies in this case are rewritten in Table 6.3b and are substituted for O in the formula. The expected frequencies (E) require a little calculation. They are found by multiplying the sum of the row by the sum of the column in which the observed frequency occurs and dividing by N, the sum of all the observed frequencies. For example, the observed frequency for r_1k_1 in Table 6.3b (the top left-hand value) is 21. The expected frequency appropriate to that number is therefore $(30)(57)/110 = 15.5$, and the fraction which will appear in the formula is $(21 - 15 \cdot 5)^2/15 \cdot 5$. (As 57 out of 110 readings were on Keuper Marl, we would expect, under H_0, $\frac{57}{110}$ of the total number of slope readings under $5°$ to be on Keuper Marl—i.e. $\frac{57}{110}$ of 30.)

TABLE 6.3b

		Keuper Marl		Greensand		
		O k_1	E	O k_2	E	Σr
r_1	0–4·8	21	15·5	9	14·4	30
r_2	5–9·8	18	15	11	13·9	29
r_3	10–14·8	9	12·4	15	11·6	24
r_4	15–19·8	8	10·9	13	10·1	21
r_5	20–24·8	1	3·1	5	2·9	6
	Σk	57		53		$N = 110$

It is now possible to substitute actual values in the formula:

$$\chi^2 = {}^r\Sigma^k\Sigma\{(O - E)^2/E\}$$

$$= \frac{(21 - 15 \cdot 5)^2}{15 \cdot 5} + \frac{(9 - 14 \cdot 4)^2}{14 \cdot 4} + \frac{(18 - 15)^2}{15} + \frac{(11 - 13 \cdot 9)^2}{13 \cdot 9}$$

$$+ \frac{(9 - 12 \cdot 4)^2}{12 \cdot 4} + \frac{(15 - 11 \cdot 6)^2}{11 \cdot 6} + \frac{(8 - 10 \cdot 9)^2}{10 \cdot 9} + \frac{(13 - 10 \cdot 1)^2}{10 \cdot 1}$$

$$+ \frac{(1 - 3 \cdot 1)^2}{3 \cdot 1} + \frac{(5 - 2 \cdot 9)^2}{2 \cdot 9}$$

$$= \underline{11 \cdot 03} \ldots$$

In order to arrive at a critical value for chi square (i.e. so that it is possible to look up the value in Appendix 4) it is necessary to calculate the appropriate degrees of freedom. In this case we take the number of rows less one and multiply the result by the number of columns less one. Or:

$$df = (r - 1)(k - 1) = (5 - 1)(2 - 1) = 4.$$

5. With 4 degrees of freedom the critical values for chi square are 9·49 at the 0·05 level and 13·28 at the 0·01 level of significance. We can therefore reject H_0 at the 0·05 level, and accept H_1, that the differences in slope angle are so great as to make it unlikely they are the result of 'chance' variations.

'Degrees of freedom' is a term that is frequently met with in statistical texts, though rarely explained. It is most easily understood by considering examples. The calculation for Table 6.2b consists of five fractions, the sum of which is 11·54. We say we have four degrees of freedom in this case because once we know the contribution of any four frequencies of individual basins the value of the fifth frequency is fixed. In other words, if the total number of tributaries is fixed at 217, then once the contribution of four basins to this total is known, the contribution of the fifth basin cannot change. We are therefore able to say that in this example $df = 5 - 1 = 4$.

TABLE 6.3c

	k_1	k_2	Σr
r_1	21	x	30
r_2	18	x	29
r_3	9	x	24
r_4	8	x	21
r_5	x	x	6
Σk	57	53	$N = 110$

The same principle applies to the calculation of degrees of freedom in the case of a contingency table. Consider the data set out in Table 6.3c (obtained from the previous example). By using the sum of k_1 we are able to establish the value of $k_1 r_5$ from the given values of $k_1 r_{1-4}$. Similarly, using the row totals, we can find all the values of $r_1 - r_5$ for k_2. Four frequencies are again the minimum number of values which enable us to determine the rest, and we say we have four degrees of freedom, or $df = (r - 1)(k - 1) = 4 \times 1 = 4$. Had Table 6.3c had a third column, eight frequencies would have been needed, giving us eight degrees of freedom, and the equation would have been $df = (r - 1)(k - 1) = 4 \times 2 = 8$.

Summary
1. Formulate H_0, and H_1. Decide upon the rejection level.
2. Cast the data in the form of Table 6.3b.
3. Find E for each value of O by multiplying the sum of that

column by the sum of that row, and dividing the product by the sum of all the observed frequencies.
4. Using Formula 6.2 calculate χ^2.
5. Obtain the significance of the result from Appendix 4.

Consolidation
1. A survey of all the farms in five small separate valleys revealed that in some of the valleys more farmers tended to concentrate on livestock than in others. Before causes were sought to explain this pattern it was decided to test the possibility that the variations observed were due to chance. The farms were of roughly comparable size. The data gathered were as follows:

Valley	A	B	C	D	E
Area	16	25	39	20	17
Number of farms with 70% income from livestock	9	7	12	3	10

Using a χ^2 test calculate the probability that this distribution is due to random variations. At what level of significance is your result?
2. A study was made of the size and number of smallholdings still worked in three of the western islands of Britain to discover whether differences existed between the three locations. The information obtained was as follows:

Size of holding in hectares	Number of holdings still worked		
	Island A	Island B	Island C
4·1 to 5	23	10	5
3·1 to 4	15	14	9
2·1 to 3	12	17	15
1·1 to 2	4	7	19
up to 1	2	1	3

There is an apparent difference in terms of size of holding between the three islands. Use a χ^2 test to determine whether this is significant, and if so, at what level.

6.3. THE BINOMIAL TEST

6.3a. The binomial test for differences of proportion

This test is used to compare two sets of dichotomized data—that is, data in which observations are assigned to one of two mutually exclusive categories—men and women, arable and non-arable fields, towns with over 50 000 people and those with less. For instance, if a map or field exercise reveals different proportions of large and small fields on slopes with northerly and southerly aspects (say, below and above 4 acres, which is about $\frac{1}{4}$ square inch on a six-inch map), the binomial test can be used to test the hypothesis that aspect and field size are related in the study area.

It is normally not possible to give a meaningful estimate of the power efficiency of the binomial test, because the test is generally used if alternatives are not available, or are of doubtful validity.

To begin with, the two sets of data (the number of small and large fields on north- and on south-facing slopes) are regarded as samples of a single population of fields; the proportion of small fields in each 'sample' is regarded as an estimate of the proportion in the total population; and any difference in sample estimates is attributed to 'chance' rather than to aspect—i.e. we have a null hypothesis that field size and aspect are unrelated. We then calculate the probability of obtaining a difference as great as that between the two sample estimates under the null hypothesis, making use of the following relationships which have already been introduced on earlier pages:

(i) The standard deviation of the sampling distribution of estimates (otherwise called the standard error of the estimate) from counted data is given by: S.E. $= \sqrt{(p\%q\%/n)}$ (Formula 5.4) where p is the percentage of the sample in a specified category; q is the percentage of the sample not in that category; n is the size of the sample; provided pqn is greater than about 9 when p and q are expressed as proportions of 1 instead of as percentages. (See p.153.) This last point needs to be stressed, because if the product of n, p, and q is less than 9 in either of the samples being compared, it means that that sample is too small for the binomial test to be valid. It is as well to check on this before proceeding with the test.

(ii) The sampling distribution of *differences* between pairs of sample estimates will tend to be normal with a standard deviation (otherwise called the standard error of the difference) given by:

$$S.E._{\text{diff.}} = \sqrt{\{(S.E._1)^2 + (S.E._2)^2\}} \qquad \textit{Formula 6.3}$$

where $S.E._1$ and $S.E._2$ are the standard errors of the two sample estimates.

If the actual difference between sample estimates is expressed as a z-score (i.e. is divided by the standard deviation of the sampling distribution of differences, otherwise called the standard error of the difference), the probability of its occurring under the null hypothesis can be found from the z-table (Appendix 1, col. B or C). The decision to retain or reject the null hypothesis can be made on the basis of that probability. Clearly, the greater the difference in the proportion of small fields on north- and south-facing slopes in our area, the smaller the probability that this difference is due to 'chance' rather than to aspect—i.e. the less likely it is that the null hypothesis is correct. Therefore, if the difference is so large that the associated probability falls below a previously determined rejection level, the null hypothesis is rejected. If it is not, then the data do not contain sufficient evidence to reject the null hypothesis, or to support the hypothesis that field size is related to aspect.

Let us now set out the formal procedure and calculations with some actual figures. In a part of east Devon, 30 out of 50 fields with a southerly aspect, and 16 out of 40 with a northerly aspect were found to have areas of less than 4 acres. (Only fields with an average gradient of over 3 degrees—50 ft contours less than one inch apart on a six-inch map—and facing either between south-east and south-west, or between north-east and north-west, were considered.)

Do these figures support the hypothesis that field size and aspect are related in the area studied? Before carrying out the test, check that npq is greater than 9 for both samples:

Southerly-aspect fields: $n = 50$ $p = 30/50 = 0.6$
$$q = 0.4$$
$$\therefore npq = 50 \times 0.6 \times 0.4 = 12$$

Northerly-aspect fields: $n = 40$ $p = 16/40 = 0.4$
$$q = 0.6$$
$$\therefore npq = 40 \times 0.4 \times 0.6 = 9.6.$$

Since both values of npq are greater than 9, we can proceed with the test:
1. Null hypothesis (H_0): field size is unrelated to aspect.
2. Alternative hypothesis (H_1): field size is related to aspect.
3. Rejection level: 0·05.
4. *The binomial test*
    (a) Calculate the percentage of small fields (p_1 and p_2) in each sample, and the difference between them, $p_1 - p_2 = d$
      (i) Southerly aspect: $p_1 = (30/50) \times 100 = 60$ per cent
      (ii) Northerly aspect: $p_2 = (16/40) \times 100 = 40$ per cent
      (iii) $d = 60 - 40 = 20$ per cent

(b) Calculate the standard error of each sample estimate of the percentage of small fields (p) and of large fields (q):

$$S.E._1 = \sqrt{(p\%_1 q\%_1/n_1)} = \sqrt{(60 \times 40/50)} = \sqrt{48} = 6{\cdot}9 \ldots$$
$$S.E._2 = \sqrt{(p\%_2 q\%_2/n_2)} = \sqrt{(40 \times 60/40)} = \sqrt{60} = 7{\cdot}7 \ldots$$

(c) Calculate the standard error of the difference:

$$S.E._{diff} = \sqrt{(S.E._1)^2 + (S.E._2)^2} = \sqrt{(48 + 60)} = \sqrt{108}$$
$$= 10{\cdot}4$$

(d) Express the actual difference between estimates as a z-score:

$$z = \frac{d}{S.E._{diff}} = \frac{20}{10{\cdot}4} = 1{\cdot}9$$

(e) Refer to z-table for the probability of obtaining a value of z at least as large as $1{\cdot}9$. In view of the fact that the alternative hypothesis (H_1) is non-directional, a two-tailed probability is required. Appendix 1 col. C gives a two-tailed probability of $0{\cdot}057$ corresponding to $z = 1{\cdot}9$.

5. The decision to reject or retain the null hypothesis: because the probability of obtaining a z-score of $1{\cdot}9$ under H_0 is more than the rejection level of $0{\cdot}05$, the null hypothesis cannot be rejected. But it is a very close thing, suggesting that slightly larger samples might well have produced a different result. As it is, we have to say that the case for the hypothesis that aspect and field size are related is not proved, and we must go away and do some more field or map work.

It is perhaps worth noting that if our alternative hypothesis, formulated before the exercise was undertaken, had been directional —i.e. that south-facing slopes had a higher proportion of small fields (for instance, because we had found more farmhouses on south-facing slopes, and we were testing the von Thünen model which associates intensive land use and therefore smaller fields with proximity to the farmhouse) then we should have required the one-tailed probability associated with a z-score of $1{\cdot}9$. This is $0{\cdot}029$ (Appendix 1 col. B) which is well below the rejection level of $0{\cdot}05$. In other words, the data collected in the exercise, when combined with previously acquired knowledge and reasonable deduction, provide sufficient evidence to uphold the hypothesis that field size and aspect are related; the data by themselves do not.

6.3b. The binomial test of differences in numbers
It is sometimes of greater interest to compare the number of observations from two samples in a given category, rather than their

proportions. For instance, we may wish to compare the number of workers in a particular industry or occupation from the sample census volumes of 1961, 1966, and 1971 (already commented upon on p.156). The procedure is exactly the same as that just outlined, except that we use Formula 5.3 which expresses the standard error of an estimate as a number instead of as a percentage:

S.E. $= \sqrt{(npq)}$

where n is the size of sample;
  p is the proportion of the sample in the given category;
  q is the proportion not in that category;
and again, where npq is greater than 9 (p.152);

  p and q are expressed as proportions of one, not as percentages.

This formula may be modified to take account of the sampling fraction, by multiplying by $\sqrt{(1-f)}$, where f is the sampling fraction (see p.155). The full formula is then:

S.E. $= \sqrt{\{npq(1-f)\}}$.

Consider the following figures, which apply to the number of men enumerated in Nottinghamshire in the 10 per cent counts for the years stated:

	1961	1966
Total economically active: (n)	297 08*	295 37
Number of miners and quarrymen:	39 30	31 92
Proportion of miners and quarrymen (p):	0·13	0·11
Proportion *not* miners and quarrymen (q):	0·87	0·89
Sampling fraction (f):	0·1	0·1

*The position of the spaces is a reminder that all numbers (but not proportions) should be multiplied by 10 to give estimates of the numbers in the *population* as opposed to the numbers counted in the 10 per cent sample, e.g. 297 08 becomes 297 080.

The question raised by these figures which the binomial test will answer is whether the difference in the two enumerations is due to there being a real decline in the number of miners and quarrymen in Nottinghamshire, or whether the difference could be accounted for by sampling errors. Again, before we begin, we should check that npq is greater than 9 in both samples. In this case, with both samples numbering nearly 30 000, it is clear that npq will far exceed 9, so we can proceed with the test.

1. Null hypothesis (H_0): both samples are taken from similar populations, and differences are due to chance in sampling.

2. Alternative hypothesis (H_1): differences are so great that the two samples are drawn from different populations.

3. Rejection level: 0·05.

4. The binomial test:

    (a) The difference in the number of miners and quarrymen (d) counted is: $3930 - 3192 = 738$.

    (b) The standard error of estimates: $\text{S.E.} = \sqrt{\{npq(1-f)\}}$

        In 1961: $\text{S.E.}_1 = \sqrt{\{29708 \times 0\cdot13 \times 0\cdot87 \times (1-0\cdot1)\}} = 55$

        In 1966: $\text{S.E.}_2 = \sqrt{\{29537 \times 0\cdot11 \times 0\cdot89 \times (1-0\cdot1)\}} = 51$

    (c) The standard error of the difference between estimates:

$$\text{S.E.}_{\text{diff}} = \sqrt{\{(\text{S.E.}_1)^2 + (\text{S.E.}_2)^2\}} \quad (Formula\ 6.3)$$
$$= \sqrt{\{55^2 + 51^2\}} = 75$$

    (d) Expressing the difference in estimates (d) as a z-score:

$$z = \frac{d}{\text{S.E.}_{\text{diff}}} = \frac{738}{75} = 9\cdot84.$$

5. Retention or rejection of null hypothesis: A two-tailed test is required, so Appendix 1 col. C is consulted. The probability associated with a z-score of 9·84 is so small that it is not even given in the table; it can be regarded as zero. In other words, the probability of obtaining a difference in sample estimates with a z-score of 9·84 under the null hypothesis is practically nil; it is certainly well below the rejection level of 0·05; one can say with almost complete confidence that the difference between the two sample figures is not due to sampling error, but to a real decline in the number of miners and quarrymen in Nottinghamshire between 1961 and 1966.

Finally, it is worth reminding readers that a quick approximation to the standard error of estimates of the number of people in a given category from the 10 per cent sample census is given by:

    the square root of that estimate multiplied by 0·9 (see p.156).

With the figures just discussed,

$$\text{S.E.}_1 = \sqrt{3930 \times 0\cdot9} = 57$$
$$\text{S.E.}_2 = \sqrt{3192 \times 0\cdot9} = 51$$

It will be seen that these are good approximations to the figures worked out in step 4(b) above, and take much less time to calculate. When fed into step 4(c), they give a standard error of the difference between estimates of 76·5 instead of 75; the z-score for the difference still approximates to 9·8, so the conclusion is exactly the same. Only in marginal cases, where the probability of a difference in estimates occurring under H_0 is very close to the rejection level, will this quick approximation prove inadequate. But it is intended specifically for 10 per cent sample census data, and should not be used with other samples without taking account of the provisos mentioned on p.156.

Summary: (for tests of both proportional and numerical differences)
1. Check that $npq > 9$ in both samples.
2. Formulate H_0 and H_1. Decide upon the rejection level.
3. Calculate the difference (d) between observed proportions or numbers in the samples.
4. Calculate the standard errors of estimates for both samples, using Formula 5.4 (for proportions) or Formula 5.3 (for numbers). If the sampling fraction (f) is known, and if it is not less than 1/10, multiply the standard errors by $\sqrt{(1-f)}$.
5. Calculate the standard error of the difference from Formula 6.3.
6. Obtain significance of the result by referring to z-tables (Appendix 1 col. B for one-tailed test; col. C for two-tailed test).

Consolidation
3. Random distributions of 50 points on each of two land-use maps yield the following results:

	Map I	Map II
No. of points on arable land	20	30
No. of points on grassland	20	15
No. of points on woodland and rough pasture	10	5

Determine, if possible, whether there is a significant difference (at the 0·05 level) between the proportion of land on the two map areas devoted to (i) arable, (ii) grass, (iii) wood and rough pasture.

4. In a 10 per cent sample census of three mining villages, the following data were obtained:

	Village A	Village B	Village C
Enumerated adult population	84	120	144
Number of miners counted	21	30	36

Calculate the significance levels of the differences between the number of miners counted in (i) villages A and B, (ii) villages B and C, (iii) villages A and C.

6.4. TESTS TO COMPARE THE MAGNITUDE OF SAMPLE MEASUREMENTS

We come now to the comparison of measured data, in which individual observations are recorded as numbers on either an interval or an ordinal scale.

A frequent problem facing the geographer is that of comparing the average magnitude (as expressed in the arithmetic mean) of two groups of observations, such as the roundness of pebble samples from different parts of a beach, the annual rainfall over successive periods, or the proportion of the work force of towns in different size categories who commute, belong to a particular industrial or occupational group, or do shift-work. In such cases, it may be required to establish the significance of the difference in samples—i.e. the likelihood that they reflect real differences in the populations they represent.

Before outlining the statistical tests by which this may be done, it is necessary to remind readers of the distinction between problems involving *paired* data, and those involving *independent* data, because they must be tested in different ways. Paired variates, it will be recalled (p.168), have something in common which they do not share with the rest of the data, so that each variate in one data set can be matched with one (and only one) variate in the other.

The importance of this distinction will be illustrated with some fictitious figures:

TABLE 6.4

Column 1 a	Column 2 b	Column 3 c	Column 4 $a - b$	Column 5 $a - c$
10	9	1	1	9
8	7	3	1	5
6	5	5	1	1
4	3	7	1	−4
2	1	9	1	−7
$\bar{a} = 6$	$\bar{b} = 5$	$\bar{c} = 5$		

Let us first suppose that a and b (columns 1 and 2 above) are *paired* samples. ($a = 10$ is paired with $b = 9$; $a = 8$ with $b = 7$, and so on.) The difference between their means ($\bar{a}$ and $\bar{b}$) is only 1, but individual values of *a* are consistently, and without exception, greater than corresponding values of *b* (column 4). Now it is a general principle in statistics that *the greater the consistentcy in a set of data, the stronger the inferences that can be made from them.* In this case, there seems good reason to believe that *a* represents a population of greater average magnitude than that represented by *b*.

Let us now suppose that *a* and *c* are paired samples. The over-all values of *c* are identical to those of *b*, but individual values are paired differently with individual values of *a*. The difference between the means of *a* and *c* will therefore be the same as that between *a* and *b*, but the relationship between individual pairs will be far less consistent (column 5), so that there cannot be the same confidence that *a* represents a population with a greater mean than that represented by *c*.

If we now think of *a* and *b* as *independent* samples, we cannot speak of the consistency of the difference between pairs of values of *a* and *b*, because they are not paired; *a* = 10 cannot be linked with any *particular* value of *b*. Consistency in the difference between pairs cannot therefore be used as a criterion in assessing the significance of the difference between samples. Clearly, problems involving independent and paired samples must be tackled differently.

We shall now present the following tests for establishing the significance of the difference in the means of two samples:

1. The *t*-test for independent samples
2. The *t*-test for paired samples } These are parametric tests

3. The *U*-test for independent samples
4. The Wilcoxon test for paired samples } These are non-parametric tests

6.5. THE *t*-TEST

Before describing the *t*-test, it should be pointed out that, in common with other parametric tests, it generally involves lengthly calculation unless samples are small and numbers easy. Also, it makes a number of assumptions which may not hold for the data to be compared; and even if they are tenable, the preliminary work necessary to demonstrate this (as described in section 6.5ciii and 6.5cv) is also time-consuming. We include the *t*-test because it is widely (and sometimes wrongly) used, and because it is the standard against which the power-efficiency of other tests is measured. In general, however, we recommend the use of non-parametric alternatives (described in sections 6.6 and 6.7) when comparing the means of two data sets—particularly to those who are not mathematically inclined.

The *t*-test makes use of the *t*-distribution already introduced in the previous chapter (p.150). It was used there to estimate confidence limits of population parameters from small samples. It will now be used to determine the significance of the difference between two groups of data measured on an interval scale. To keep the arithmetic simple, we shall again invent some figures, and use them first as independent samples, and then as paired samples, thereby illustrating two quite different problems with very different results.

6.5a. The t-test for independent samples

Sample values of two independent variables, x and y, together with deviations from sample means (($x - \bar{x}$) and ($y - \bar{y}$)), and the squares of these deviations, are given in Table 6.5.

TABLE 6.5

x	y	$(x - \bar{x})$	$(x - \bar{x})^2$	$(y - \bar{y})$	$(y - \bar{y})^2$
10	8	3	9	2	4
8	7	1	1	1	1
7	7	0	0	1	1
6	6	−1	1	0	0
6	5	−1	1	−1	1
5	3	−2	4	−3	9
$\bar{x} = 7$	$\bar{y} = 6$		$\Sigma(x - \bar{x})^2 = 16$		$\Sigma(y - \bar{y})^2 = 16$

$n_x = 6 = n_y$ = size of each sample (but samples do not need to be equal in size for a t-test for independent samples).

Before the t-test can be applied, two assumptions must be made:
- (i) The background populations of the samples are approximately normally distributed. This is particularly important in this example, because the samples are so small (see p.170).
- (ii) The standard deviations of the populations from which the samples are drawn are equal.

Checks to determine whether these assumptions are tenable are deferred until later in the chapter to avoid complicating the exposition, but they should really be carried out before the t-test is used.

We shall now follow the steps in hypothesis testing outlined on p.160:

1. Null hypothesis (H_0): there is no difference between the means of the populations of which x and y are samples.

2. Alternative hypothesis (H_1): there is a difference between the means of the populations of which x and y are samples.

3. Rejection level (α): 0·05

4. *The test*:

(a) Calculate the standard deviation of the populations. Because of assumption (ii) above, and because the samples are so small, a *pooled best estimate*[4] ($\hat{\sigma}$) is obtained from all the data in both samples, from the formula:

$$= \sqrt{\left(\frac{\Sigma(x-\bar{x})^2 + \Sigma(y-\bar{y})^2}{n_x + n_y - 2} \right)}$$

Formula 6.4

$$= \sqrt{\left(\frac{16+16}{6+6-2} \right)} = \sqrt{3\cdot2}$$

(b) Calculate the standard deviations of the sampling distributions of $\bar{x}$ and $\bar{y}$ (i.e. the standard errors of sample means):

$$\text{S.E.}_{\bar{x}} = \hat{\sigma}/\sqrt{n_x} = \sqrt{3\cdot2}/\sqrt{6} = \sqrt{0\cdot53} = 0\cdot73 \qquad \text{(using}$$
$$\text{S.E.}_{\bar{y}} = \hat{\sigma}/\sqrt{n_y} = \sqrt{3\cdot2}/\sqrt{6} = \sqrt{0\cdot53} = 0\cdot73 \quad \text{Formula 5.1)}$$

(c) Calculate the standard deviation of the sampling distribution of the difference-between-means (i.e. the standard error of $(\bar{x}-\bar{y})$). This is obtained, as in the binomial test, by the formula:

$$\text{S.E.}_{\cdot(\bar{x}-\bar{y})} = \sqrt{\{(\text{S.E.}_{\bar{x}})^2 + (\text{S.E.}_{\bar{y}})^2\}} \qquad \text{(using Formula 6.3)}$$

Therefore,

$$\text{S.E.}_{\cdot(\bar{x}-\bar{y})} = \sqrt{\{0\cdot53 + 0\cdot53\}} = \sqrt{1\cdot06} = 1\cdot03$$

(d) Calculate t:

$$t = \frac{\text{the difference between the means}}{\text{the standard error of the difference}} \qquad \textit{Formula 6.5}$$

$$t = \frac{\bar{x}-\bar{y}}{\text{S.E.}_{\cdot(\bar{x}-\bar{y})}} = \frac{6-5}{1\cdot03}, \text{ which is approximately 1.}$$

(e) As the rejection level is 0·05, enter the t-table (p.332) under the column-heading 0·05, and along the row for ($n_x + n_y - 2$) degrees of freedom—i.e. 10 degrees of freedom. This gives a critical value for t of 2·23. For the difference between sample means ($\bar{x}-\bar{y}$) to be significant, t must be more than 2·23. (Note: These are two-tailed critical values of t, which is appropriate for this example, because H_1 was non-directional. If H_1 had been directional, the critical value of t would have been found under the 0·10 column heading, since a two-tailed probability of 0·10 is equal to a one-tailed probability of 0·05. The critical value would then have been 1·81.)

5. The decision to reject or retain H_0: For H_0 to be rejected, t must be more than the critical value found in the t-table. In this case, t was

calculated to be 1; the critical value was found to be 2·23; therefore H_0 cannot be rejected; the difference between $\bar{x}$ and $\bar{y}$ is not significant at the 0·05 level.

6.5b. *The t-test for paired samples*

The same values of x and y are used to illustrate this method, but because individual values of x and y are paired, the differences between them (d) are used:

TABLE 6.6

x	y	d	$(d - \bar{d})$	$(d - \bar{d})^2$	
10	8	2	1	1	
8	7	1	0	0	
7	7	0	−1	1	Number of pairs
6	6	0	−1	1	$= n = 6$.
6	5	1	0	0	
5	3	2	1	1	
$\bar{x} = 7$	$\bar{y} = 6$	$\bar{d} = 1$		$\Sigma(d - \bar{d})^2 = 4$	

The assumptions and steps 1 to 3 outlined for the t-test for independent samples apply again here. The difference in procedure comes in step 4, *the test*:

(a) Calculate the best estimate of the standard deviation of the difference (d) between paired values of x and y, using Formula 5.2b

$$\hat{\sigma}_d = \sqrt{\left\{\frac{\Sigma(d - \bar{d})^2}{n - 1}\right\}} = \sqrt{\left(\frac{4}{5}\right)} = \sqrt{0 \cdot 8}$$

(b) Calculate the standard deviation of the sampling distribution of mean differences (i.e. the standard error of mean differences):

$$\text{S.E.}_{\bar{d}} = \hat{\sigma}_d / \sqrt{n} = \sqrt{0 \cdot 8} / \sqrt{6 \cdot 0} = \sqrt{0 \cdot 13} = 0 \cdot 36$$

(using Formula 5.1)

(c) Calculate t:

$$t = \frac{\text{the difference between the means}}{\text{the standard error of the difference}} \qquad \text{(Formula 6.5)}$$

$$t = \frac{\bar{d}}{\text{S.E.}_{\bar{d}}} = \frac{1}{0 \cdot 36} = 2 \cdot 8$$

(d) As the rejection level is again 0·05, and as H_1 is again non-directional, enter the t-tables (which give two-tailed probabilities) in the column headed 0·05, and opposite ($n - 1$) degrees of freedom—

i.e. opposite 5 degrees of freedom. The critical value of t is seen to be 2·57.

Step 5: The decision to reject or retain H_0. For H_0 to be rejected, t must be more than the critical value found in the t-table. In this case, t was calculated to be 2·8; the critical value of t is 2·57; therefore H_0 can be rejected; the difference between $\bar{x}$ and $\bar{y}$ is significant at the 0·05 level.

We see then that if values of x and y are independent (as in Table 6.5), their means are not significantly different; if the same values are paired (as in Table 6.6), their means are significantly different.

TABLE 6.7
Annual rate of population change 1951–61, and 1961–71 in samples of large and small towns (with populations of 150 000 to 300 000 and 50 000 to 100 000 in 1971 respectively):

	Large Towns 1951–61	1961–71	Small Towns	1961–71
	x	y		z
Nottingham	0·13	−0·40	Preston	−1·51
Leicester	0·08	−0·16	Darlington	0·16
Stoke	−0·09	−0·44	Barnsley	0·08
Derby	0·81	0·31	Hemel Hempstead	2·30
Portsmouth	−0·82	−0·87	Cheadle and Hulme	2·89
Swansea	0·39	0·31	Cannock	1·85
Warley	−0·58	−0·41	St. Albans	0·35
Bolton	−0·39	−0·43	Margate	0·92

Note: The sample was selected and the data taken from Table 6 of the Preliminary Report of the 1971 Census, in which towns are listed in order of size; the towns named above were selected at regular intervals from this list.

Note also: The following calculations, based on the above table, will prove useful in answering the Consolidation Exercise:

$$\bar{x} = -0.06 \quad \Sigma(x-\bar{x})^2 = 1.943$$
$$\bar{y}' = -0.36 \quad \Sigma(y-\bar{y})^2 = 1.213$$
$$\bar{z} = +0.88 \quad \Sigma(z-\bar{z})^2 = 14.150$$
$$\Sigma(x-y)^2 = 0.750$$

Consolidation
5. Refer to Table 6.7. Use t-tests to determine whether:

(a) Large towns and small towns in England and Wales (as defined in the table) experienced significantly different rates of population change during the period between the censuses of 1961 and 1971.

(b) Rates of change of population in large towns were significantly different during the periods 1951–61 and 1961–71. (Note: The results of some of the more lengthy calculations are given above.)

6.5c. Additional observations on the t-test
Before leaving the *t*-test, some important additional observations are necessary.

6.5ci. Large samples
The *t*-distribution is generally used to compare the means of small samples. It can, however, be used for large samples, applying the same procedures as those already outlined. But if the *t*-table is examined and compared with the *z*-table (for a normal distribution), it will be observed that: (a) critical values of *t* do not vary much with degrees of freedom if the latter are over about 40; (b) the probabilities associated with the *t*-distribution with more than 40 degrees of freedom are very similar to those associated with a normal distribution (which are not dependent on degrees of freedom); therefore, the *t*-distribution for large samples is very similar to the normal distribution.

It follows that, for *independent* samples where $n_x + n_y$ exceeds 40 (so that degrees of freedom = $n_x + n_y - 2$ exceed 38); and for *paired* samples with over 40 pairs (degrees of freedom = $n - 1$ = more than 39), *z*-tables can be used just as well as *t*-tables, and degrees of freedom can be ignored.

6.5cii. Power-efficiency
The *t*-tests are the most powerful tests available for the comparison of sample means. Their power-efficiency is therefore 100 per cent, and they are the standard against which the power-efficiency of non-parametric alternatives such as the Mann-Whitney *U* and the Wilcoxon tests is measured (p.168).

6.5ciii. Checks for normality
It has been pointed out that the *t*-test is only valid if the background populations of samples are approximately normal. Two checks or criteria can be applied:

(i) Common sense and some knowledge of the factors affecting the variables in question—are the data of a kind one might expect to cluster symmetrically about a mean, for reasons described in Chapter Four (p.120)?

(ii) The use of probability paper—if the sample data are plotted cumulatively (as explained on p.127), they will yield a straight line if they are normally distributed.

6.5civ. Transformation of skewed distributions   If samples are not normally distributed, it may be possible to transform the data so that they are more nearly normal and therefore amenable to the *t*-test (and other parametric tests). Many distributions have a single mode like the normal distribution, but are asymmetrical, being either positively or negatively skewed as in Figure 6.4. Examples of the former may be the number of people (or countries) in different income brackets and towns in different size categories. Negatively skewed distributions are less common. An example may be the percentage of native-born people in the populations of towns, regions, or countries. The most widely used transformation for positively skewed distributions is that in which numbers are replaced by their logarithms. Below are a set of numbers (top line), together with their logarithms correct to two decimal places (bottom line):

Nos.	12	15	18	20	22	24	26	30	32	34	36
Logs.	1·08	1·18	1·26	1·30	1·34	1·38	1·42	1·48 s	1·51 s	1·53	1·57

Nos.	38	44	48	52	56	68	81
Logs.	1·58	1·64	1·68	1·72	1·75	1·83	1·91

If the *numbers* are grouped, a positively skewed frequency distribution results as in Figure 6.5a; if the logarithms are grouped, the resulting distribution is symmetrical and more nearly normal (Figure 6.5b).

Therefore, by substituting the logarithms for the numbers, we can transform this skewed distribution into one which is approximately normal and therefore satisfies the requirements of parametric tests such as the *t*-test. Of course, if two samples are being compared, they must be transformed in exactly the same way. Provided this is done, the results of any hypothesis test carried out on the transformed values holds good for the original values.

A negatively skewed distribution can be turned into a positively skewed distribution, so that a logarithmic transformation can then normalize it. This is done by subtracting each value from a number

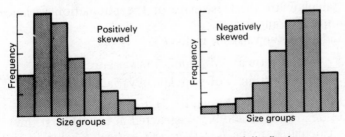

Figure 6.4. Positively and negatively skewed distributions

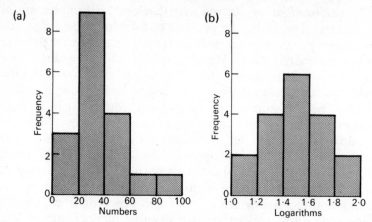

Figure 6.5. The effect of a logarithmic transformation on a positively skewed distribution

greater than the highest value. For instance, the following series, when grouped, is seen to be negatively skewed (Figure 6.6a):

1, 2, 3, 4, 4, 5, 5, 6, 6, 7, 7, 7, 8, 8, 8, 8, 9, 9, 10

If each value is subtracted from 11, the series becomes:

10, 9, 8, 7, 7, 6, 6, 5, 5, 4, 4, 4, 3, 3, 3, 3, 2, 2, 1.

If these figures are now grouped, a positive skew results (Figure 6.6b) so that if they are now transformed logarithmically, the distribution will be nearer normal.

6.5cv. The variance ratio test The validity of a *t*-test also rests on the assumption that the standard deviations of the background popu- lations of the two samples are equal. We should therefore check that the best estimates of the population standard deviations derived from the two samples taken separately are not so different as to render the assumption unacceptable. This can be done by applying the variance ratio test (variance is the square of the standard deviation—see p.19) as follows:

(a) Calculate the best estimate of the population variance from each sample ($\hat{\sigma}_1^2$ and $\hat{\sigma}_2^2$).

(b) Calculate the variance ratio (F):

$$F = \frac{\text{greater estimate of population variance } (\hat{\sigma}_1^2)}{\text{smaller estimate of population variance } (\hat{\sigma}_2^2)} \quad \textit{Formula 6.6}$$

(c) Refer to F table (Appendix 5). Enter the table at $(n_1 - 1)$ and $(n_2 - 1)$ degrees of freedom as instructed in the table. If the variance ratio is calculated to be *less* than the critical F-value in the table, then the difference between the best estimates of the population

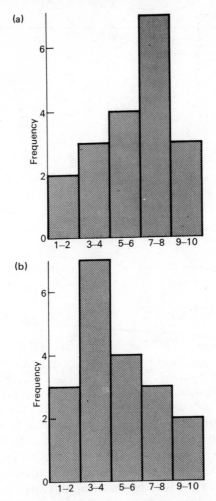

Figure 6.6. Converting a negatively into a positively skewed distribution

variances (and therefore standard deviations) based on the two samples is not so great as to be incompatible with the assumption that the standard deviations of the two populations are equal. The t-test can then be carried out. If the variance ratio is found to be greater than the critical value, then the necessary assumption of a common population standard deviation must be regarded as inconsistent with the data, and the t-test cannot be used.

Example   Table 6.8 sets out sample values of variables a and b, together with some necessary calculations for carrying out the variance ratio test on the two samples:

TABLE 6.8

a	b	$(a-\bar{a})$	$(a-\bar{a})^2$	$(b-\bar{b})$	$(b-\bar{b})^2$
10	7	5	25	2	4
8	6	3	9	1	1
6	5	1	1	0	0
4	4	−1	1	−1	1
2	3	−3	9	−2	4
0		−5	25		

$\bar{b} = 5$

$\bar{a} = 5$ $\Sigma(a-\bar{a})^2 = 70$ $\Sigma(b-\bar{b})^2 = 10$

$n_b = 5$

$n_a = 6$

Applying the variance ratio test:

(a) $\hat{\sigma}_a^2 = \dfrac{\Sigma(a-\bar{a})^2}{n_a - 1} = \dfrac{70}{5} = 14$ (from Formula 5.2b)

$\hat{\sigma}_b^2 = \dfrac{\Sigma(b-\bar{b})^2}{n_b - 1} = \dfrac{10}{4} = 2\cdot5$

(b) Variance ratio $= 14/2\cdot5 = 5\cdot6$ (using Formula 6.6)

(c) Refer to F table. These are given for the 0·05 significance levels.

The number of degrees of freedom for larger estimates is $6 - 1 = 5$ and the number of degrees of freedom for smaller estimates is $5 - 1 = 4$. The critical value of F with these degrees of freedom is 6·3. The calculated value of F is less than this, so that the assumption of an equal population variance (and standard deviation) for the two populations represented by these samples cannot be rejected at the 0·05 level and can therefore be regarded as acceptable (meaning that a variance ratio as high as this would be expected from a common population variance more than 5 times in a hundred, which is by no means improbable).

Concluding comment: It will be observed that, in order to use the *t*-test, a good deal of preparatory checking of assumptions is required. On the other hand, the assumptions may appear more restrictive than in fact they are. They do not have to be proved to be absolutely correct; rather, they must be proved not to be incorrect. In the case just worked through, a variance ratio of 5·6 (which means that the standard deviation estimate from one sample is more than twice that from the other) is not too great to invalidate the assumption. Furthermore, it is more power-efficient than any alternative non-parametric test, meaning that somewhat smaller samples are required to achieve the same level of discrimination. Nevertheless, the checking is time-consuming; in addition, as has already been pointed out, the arithmetic involved in a t-test is not generally as easy as in the contrived examples given. For these reasons, the non-parametric alternatives to the t-test are generally to be preferred, and these will be described in the following sections.

Summaries of t-*tests*

A. *For independent samples*:
1. Formulate H_0 and H_1. Decide upon rejection level.
2. Check for normal distribution and equality of variance (sections 6.5ciii and v).
3. Calculate sample means and the difference between them $(\bar{x} - \bar{y})$ or $\bar{d}$.
4. Calculate pooled best estimate of standard deviation (Formula 6.4).
5. Calculate the standard error of means (Formula 5.1) and standard error of difference between means (Formula 6.3).
6. Calculate t (Formula 6.5).
7. Obtain significance of result from Appendix 3.

B. *For paired samples*:
1. ⎫
2. ⎬ As above
3. ⎭
4. Calculate best estimate of standard deviation of differences between paired variables (Formula 5.2b)
5. Calculate standard error of mean difference between samples $(S.E._{\bar{d}})$
6. ⎫ As above
7. ⎭

6.6. THE MANN-WHITNEY U TEST

This is one of the most powerful distribution-free tests. Even when only medium-sized samples (i.e. between 10 and 20) are involved it has nearly 95 per cent of the power of Student's 't'-test (see p.168) when used with data that conform to 't'-test requirements. It can be applied to small, medium, and large samples. Because it is a non-parametric test no assumptions need be made about the characteristics of the distribution of the populations concerned. It can be employed when only ordinal measurement is possible so long as both data sets are ranked in a single sequence, or when measurements on an interval scale are allotted rank numbers in a single sequence. It is used to test whether a difference in the mean of two independent samples is statistically significant, that is that the samples come from different populations. The test is particularly useful to the geographer when the assumptions necessary for the 't'-test cannot be safely inferred, when only ordinal measurements is possible, or to avoid the rather lengthy calculations that the 't'-test involves.

As part of a survey to investigate the influence of Douglas (I.o.M.) as the principal urban centre on the island it was found that a high proportion of people came to Douglas from the other three urban centres for their weekly shopping on the assumption that prices were lower. To check this belief a 'basket' of common household requisites was priced at six randomly chosen stores in Douglas, and two in each of the smaller towns. The resulting data are given in Table 6.9.

TABLE 6.9

Prince in pence of a 'basket' of common household requisites sold in retail shops in four towns on the Isle of Man

n_1	Area A	The three small towns	135	137	144	140	146	145
n_2	Area B	Douglas	132	133	142	134	136	143

Both samples in this case are of the same size, $n_1 = n_2 = 6$, but it is not a requirement of the test. When samples are of different sizes the smaller of the two is termed n_1.

Retail prices result from the decision of individual store keepers. They are based upon a number of considerations including availability, convenience of the location of the shop for the customer, demand, and so on. The mean price in Area B was 136·67 p, and in Area A was 141·17 p, so it appeared that shopping in Douglas was cheaper than in the smaller towns. At the same time it was recognized some shops in the smaller towns provided cheaper goods than the more expensive ones in Douglas. The question was to determine from this small sample whether the difference in the mean of the two distributions of prices was statistically significant (i.e. that they were samples from different populations).

1. The null hypothesis, H_0, was the prices in A and B form part of the same distribution and that differences between them were the result of 'chance' variations, and therefore not significant.

2. The alternative hypothesis, H_1, was that the price distribution in B was significantly lower than that in A.

3. Level of rejection of H_0 was decided at $\alpha = 0.05$. The test was directional (and therefore one-tailed), because most of the larger stores in Douglas, and especially the 'self-service' variety, were designed for a larger turnover in order to be competitive with shops in the smaller towns, where their kind of scale economies were not possible.

4. To apply the test all the values obtained must be placed in rank order, but the group identity of each value retained. Thus:

B	B	B	A	B	A	A	B	B	A	A	A
132	133	134	135	136	137	140	142	143	144	145	146

Let the number in group A be n_1, and in group B be n_2.

U is the statistic required and is obtained by inspecting each B in turn and counting the number of A's which precede it. Thus the first three B's above have no A preceding them, but the fourth B (136) has one A, and the fifth B (142) has three preceding A's. The sixth B (143) also has three preceding A's. In this case therefore $U = 0 + 0 + 0 + 1 + 3 + 3 = 7$.

Appendix 6 (A to F) gives the probabilities associated with the calculated value of U for values of n_2 from 3 to 8. With $U = 7$ and $n_1 = n_2 = 6$ it will be seen that the one-tailed probability of a difference as great as or greater than that between the two distributions occurring by 'chance' is 0.024 (i.e. half 0.047). Our rejection level was 0.05, so we can reject H_0 and accept H_1 that the prices represented by sample B were significantly lower than those in sample A and there was a 95 per cent probability that these were not the result of random variations.

It will be seen that if the number of times a B value precedes an A value is counted an inflated U results, which is termed U_1. This may happen, in which case the U_1 value will not be found in Appendix 6, because it is the lower value which is required. In our example, counting the B scores preceding an A we find:

$$U_1 = 3 + 4 + 4 + 6 + 6 + 6$$
$$= 29$$

If this occurs U_1 may be transformed to U by the formula:

$$U = n_1 n_2 - U_1 \qquad \textit{Formula 6.7}$$

In our case with $n_1 = 6 = n_2$
$$U = n_1 n_2 - U_1$$

$$= 6 \times 6 - 29$$
$$= 36 - 29$$
$= 7$, which is the value already counted for U above.

Summary
For small samples n_1 and $n_2 < 9$.

1. Formulate H_0 and H_1 and decide on the rejection level.
2. Place all the values (or ranks), in a single rank order, retaining the group identity of each. Let one group be termed A and the other group B.
3. Obtain the value of U by inspecting each B rank in turn and counting the number of A's which precede it.
4. Apply Formula 6.7 as a check.
5. Determine the significance of U from Appendix 6.

The method used above was presented to give some understanding of how the test worked, but it is also possible to find the value of U from a formula which is commonly used as the size of sample increases. This is

$$U = n_1 n_2 + \tfrac{1}{2} n_1 (n_1 + 1) - R_1 \qquad \textit{Formula 6.8a}$$

or $\quad U = n_1 n_2 + \tfrac{1}{2} n_2 (n_2 + 1) - R_2 \qquad \textit{Formula 6.8b}$

where $R_1 =$ the sum of the ranks given to values in n_1, and $R_2 =$ the sum of the ranks given to values in n_2.

Let us take the figures for mean wheat yields in cwt. per acre in the year 1967/8 for the counties of Wales, and counties in England in the East Midlands and East Anglia, given below in Table 6.10, and allot a rank for each figure. Conventionally the smallest value is given the rank of 1. If the largest value receives the rank of 1 the values of U are reversed.

Inspection of the data reveals no startling difference between England and Wales. (Remember we are dealing with *yield*, not quantity produced.) The mean yield in E. England is 33·96 cwt. per acre, and that for Wales is 31·36. The highest yield in Wales (36·4) was comparable with that in E. England (36·6), and the lowest yield in Wales (28·2) was not very much lower than the lowest in E. England (29·8). Is it possible to draw any inference from these figures? First it is necessary to formulate a hypothesis to test whether the variations observed in the two groups are likely to be the result of 'chance'.

TABLE 6.10
Wheat yield in cwt. per acre in counties in Wales, and in the East Midlands and E. England

England $n_1 = 10$		Rank	Wales $n_2 = 13$		Rank
Leicestershire	32·8	13	Anglesey	30·5	6
Derbyshire	31·9	11	Brecon	28·6	2
Lincs. Holland	36·6	23	Caernarvonshire	31·6	9
Lincs. Kesteven	35·6	20	Cardiganshire	31·7	10
Lincs. Lindsey	36·1	21	Carmarthen	32·0	12
Northamptonshire	29·8	4	Denbighshire	33·2	15
Nottinghamshire	34·5	18	Flintshire	29·6	3
Rutland	34·0	17	Glamorgan	32·9	14
Norfolk	34·9	19	Merioneth	31·5	8
Suffolk	33·4	16	Monmouth	30·1	5
			Montgomery	36·4	22
			Pembrokeshire	31·4	7
			Radnor	28·2	1
	$\Sigma R_1 = 162$			$\Sigma R_2 = 114$	

1. H_0, therefore, states that there is no significant difference between the means of the two groups.

2. If H_0 is rejected, then H_1 states that the values within the two groups do not form part of the same population, and we are entitled to conclude that the mean wheat yield is significantly greater in E. England than in Wales.

3. Our rejection level is decided at $\alpha = 0.05$. We are testing here whether yield is *greater* in one area than another, because we know that in general E. England is more suitable for wheat growing with higher yield expected. The test is therefore directional and one-tailed.

4. Applying the data given in Table 6.10 to Formula 6.8a,

$$U = n_1 n_2 + \tfrac{1}{2} n_1 (n_1 + 1) - R_1$$
$$= 10 \times 13 + \tfrac{1}{2} \times 10(10 + 1) - 162$$
$$= 23$$

If we use Formula 6.8b the result is

$$U = n_1 n_2 + \tfrac{1}{2} n_2 (n_2 + 1) - R_2$$
$$= 10 \times 13 + \tfrac{1}{2} \times 13(13 + 1) - 114$$
$$= 107$$

107 is the 'inflated' U value, U_1, referred to above. This can be proved (and our arithmetic checked) by applying Formula 6.7.

$$U = n_1 n_2 - U_1$$
$$= 10 \times 13 - 107$$
$$= 23, \text{ which is the } U \text{ value found initially.}$$

5. We now refer to Appendix 6 (G to J), which contains the critical values of U for sample sizes n_1 1 to 20, and n_2 9 to 20. It will be noted that these are not exact probabilities. For the result to be at the level of significance stated at the head of the table the calculated value of U must be equal to or *less* than the critical value given in the table for the appropriate sizes of n_1 and n_2. Note that U differs from most tests in that a value *lower* than the critical value is required. For χ^2, z-scores, and 't', a *higher* value is necessary.

The critical value for U at $\alpha = 0 \cdot 05$ (one-tailed), and $n_1 = 10$, and $n_2 = 13$, is 37. We may therefore reject H_0 at the stipulated level of significance. But it will be observed that our calculated value of $U = 23$ is also significant at the $0 \cdot 01$ level (one-tailed).

The degree of confidence in the rejection of H_0 is surprising if one refers casually to the data in Table 6.10, and recalls that the mean values for each group were relatively close. It is a good example of the danger, not of drawing conclusions from untested evidence (a mistake geographers have so frequently made in the past), but of over-looking what appears to be a highly significant statistical relationship.

Summary
For medium-size samples, when $n_1 \leqslant 20$ and n_2 is between 9 and 20. Proceed as for small samples in steps 1 and 2. Then

3. Allot ranks to each of the values, with both groups in a single sequence.
4. Sum the ranks of group A (or alternatively group B).
5. Obtain the value of U from Formula 6.8.
6. Determine the significance of the result from Appendix 6.

With *large samples* of more than 20 it is not possible to use either of the tables previously employed. However, because the sampling distribution of U approaches normal as sample size increases it is possible to establish the probability of an occurrence by use of a z-score (p.43), based on the normal distribution. The z number for the U test is found from the formula

$$z = \frac{U - \frac{1}{2} n_1 n_2}{\sqrt{\{(n_1)(n_2)(n_1 + n_2 + 1)/12\}}} \qquad \textit{Formula 6.9}$$

U may be found from either Formula 6.8a or 6.8b above, because the value of z will be the same in either case, although the sign will be different.

In a geomorphological study of the characteristics of the River Noe at its junction with Grinds Brook two random samples of stones were taken from near the mouth of Grinds Brook, and from a river gravel bank slightly downstream. The object was to determine whether the bank at this point was fed by stones brought down by the river itself or from the Grinds Brook.

1. Our null hypothesis, H_0, was that there was no significant difference between the two samples in terms of stone roundness.

2. H_1 stated that the pebbles in the main stream of the river did not form part of the same population as those from Grinds Brook, a two-tailed test because we have no means of knowing which sample should have the more rounded stones.

3. Rejection level was decided at $\alpha = 0.01$.

4. The roundness coefficients of the stones from the two samples were measured in the usual way and found to be as recorded in Table 6.11.

TABLE 6.11
Stone roundness coefficients of samples from Grinds Brook at its junction with the River Noe

$n_1 = 20$	51	32	23	6	13	13	19	
	47	27	20	7	12	13	19	
	39	27	24	9	11	17		
$n_2 = 23$	58	10	52	47	23	27	38	30
	57	16	47	45	25	28	33	30
	56	19	47	44	27	29	30	

Each value must first be ranked as before, the lowest receiving the rank of 1. The ranks are allotted in respect of *all* the data, but the identity of each sample retained (see Table 6.12). Where equal values occur the mean rank value is given to each. Thus in n_1 three stones have a coefficient of 13 and share the rank of 8, being the mean of 7, 8, and 9. Ranks 7 and 9 are omitted, and 16, the next value above 13 (in n_2), given the rank of 10.

The value of U was found from Formula 6.8.

$$U = n_1 n_2 + \tfrac{1}{2} n_1 (n_1 + 1) - R_1$$
$$= (20 \times 23) + \tfrac{1}{2} \times 20(20 + 1) - 307$$
$$= 363$$

TABLE 6.12

Stone roundness coefficients of samples from Grinds Brook at its junction with the River Noe

n_1	R_1	n_2	R_2
51	39	58	43
47	36·5	57	42
39	32	56	41
32	29	52	40
27	21·5	47	36·5
27	21·5	47	36·5
24	18	47	36·5
23	16·5	45	34
20	15	44	33
19	13	38	31
19	13	33	30
17	11	30	27
13	8	30	27
13	8	30	27
13	8	29	25
12	6	28	24
11	5	27	21·5
9	3	27	21·5
7	2	25	19
6	1	23	16·5
	307·0	19	13
		16	10
		10	4
			639·0

The z value was then calculated using Formula 6.9.

$$z = \frac{U - \frac{1}{2}n_1 n_2}{\sqrt{\{(n_1)(n_2)(n_1 + n_2 + 1)/12\}}}$$

$$= \frac{363 - \frac{1}{2} \times 20 \times 23}{\sqrt{\{(20)(23)(20 + 23 + 1)/12\}}}$$

$$= \frac{133}{\sqrt{1687}}$$

$$= 3\cdot2 \ldots$$

5. Reference to Appendix 1C shows that the probability associated with a z-value of 3·2 is 0·001. The null hypothesis can therefore be rejected at the highest level of confidence. The acceptance of H_1 indicated that the pebbles forming the bank in the main stream were supplied from higher up the main river channel, and not from the

Grinds Brook. Having established this fact, the study of the characteristics of the river and its tributary could proceed on a sound statistical basis.

Summary
For large samples proceed as for medium-size samples in steps 1 to 5.
6. Find a z-score for the value of U obtained in 5.
7. Determine the significance of the result from Appendix 1.

Consolidation
6. From the data given in Table 6.7 use the Mann-Whitney test to determine whether large towns and small towns had significantly different rates of population change between the years 1961 and 1971. (The result may be compared with that obtained by using the 't'-test for the question on p.193.)

6.7. THE WILCOXON TEST FOR PAIRED SAMPLES
The Wilcoxon test for paired samples (let us call them A and B), although used to compare data on an interval scale, is a nonparametric test in that the differences between pairs of data from the two samples are ranked—i.e. are put on an ordinal scale—and significance levels depend upon the allocation of these ranks between those cases where A is greater than B, and those where B is greater than A. By replacing numerical differences by ranks, some information contained in the data is lost (differences of adjacent rank will differ by 1, whatever their numerical size). But by so doing, the test is made 'distribution-free'—i.e. its validity does not depend upon the interval-scale values in the two samples being distributed normally or in any other special way, as is the case with the t-test. Like Mann-Whitney, it has a power-efficiency of 95 per cent (see p.168).

There are two ways of carrying out the Wilcoxon test, for small and for large samples respectively. They are identical in the earlier stages, but differ in the final step. The small samples method (for samples up to 25) will be outlined first, and then the different procedure in the final step for large samples will be described.

Table 6.13 contains data taken from $2\frac{1}{2}$-inch map No. TL01. A comparison has been made between the NE. and SW. slopes on opposing sides of a short stretch of the Ver valley, on the dip slope of the Chilterns, to determine whether there is any significant difference between their mean slopes. Measurements of contour spacing along cross-profiles have been made at regular intervals, and are used as surrogates for slope values. The figures in the first two columns (A and B) of the table give the number of millimetres on

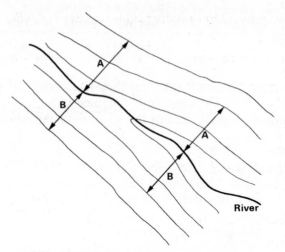

Figure 6.7. Diagram to explain data in Table 6.13

the map from the valley axis to the third contour above it, as illustrated in Figure 6.7, so that the greater the values of A and B, the
more widely spaced the contours, and therefore the less steep the
slopes represented. As each value of A can be matched with a value
of B (opposing slopes on the same cross-profile), a test for paired
samples rather than for independent samples is appropriate.
As with every other test described in this chapter, the Wilcoxon test
begins by assuming a null hypothesis—that the two sets of data (A
and B) come from identical populations (i.e. the mean gradients of
the NE. and SW. slopes of the valley are equal); and then determines
the probability of these data occurring by 'chance' if this assumption
is correct. If the probability is found to be very slight, then the
assumption is unlikely to be correct and is rejected; so that an alternative hypothesis which follows logically from the rejection of the null
hypothesis is adopted instead.

The procedure will now be set out formally:

1. Null hypothesis (H_0): the mean gradients of the NE. and SW.
slopes could be equal.

2. Alternative hypothesis (H_1): the mean gradients of the NE. and
SW. slopes are unequal. This requires therefore a two-tailed test,
since we do not specify which slope is steeper.

3. Rejection level (α): 0·05 (i.e. if the probability of these data
occurring by chance under H_0 is less than 1 in 20, then H_0 is unlikely
to be correct and is rejected).

TABLE 6.13
The contour interval of opposing slopes of the Ver valley

A 1	B 2	$\|A-B\|$ 3	Ranks of $\|A-B\|$ 4	Assigned Ranks $R_1(A>B)$ 5	$R_2(A<B)$ 6
10	11	1	1		1
7	7	0	dropped from sample because $A=B$		
9	13	4	6		6
10	8	2	2·5	2·5	
15	12	3	4·5	4·5	
12	15	3	4·5		4·5
8	10	2	2·5		2·5
9	15	6	7		7
				$R_1=7·0$	$R_2=21$

4. The test:
(a) Find $|A-B|$, the numerical difference between matched pairs, irrespective of whether A is more or less than B (column 3 of Table 6.13). Drop from the sample any pairs where $|A-B|=0$.
(b) Rank the values of $|A-B|$ with the smallest value ranked 1 (col. 4). Where values tie, assign average rank to each value, as explained on p.205.
(c) Put ranks into two columns:
R_1–those for pairs where A exceeds B (col. 5)
R_2–those for pairs where B exceeds A (col. 6).
(d) Add columns 5 and 6 (R_1 and R_2). Whichever total is the *smaller*, call T (in this case, $R_1=7·0=T$).
Before describing the final step, a word of explanation will be helpful. If H_0 is correct, R_1 and R_2 will be about equal, since the difference between paired values will be distributed randomly between profiles in which A is greater than B, and those in which B is greater than A. Therefore, the greater the difference between R_1 and R_2, the more unlikely it is that H_0 is correct. Further, since (R_1+R_2) is constant for a sample of given size n (being the sum of all the ranks from 1 to n, which can be shown to be $\frac{1}{2}n(n+1)$), the greater the difference between R_1 and R_2, the less will be the smaller value (which we have called T) and the greater will be the other. *It follows that the less the value of T, the less likely it is that H_0 is correct.*
5. *Decision to reject or retain H_0*
Refer to Appendix 7 which gives critical values of T for samples of different size from $n=6$ to $n=25$. (The Wilcoxon test cannot be used with samples of less than 6.) As the rejection level has been set at 0·05, and as the alternative hypothesis is non-directional, use the

column headed 0·05 for two-tailed tests. If the calculated value of T is *less* than the critical value for the size of sample being tested, then the difference between the two samples is significant at the 0·05 level, and H_0 is rejected. In the example given, however, the critical value is given in the table as 2 (n being 7 after one pair of values has been dropped from the sample because they are equal), and T was calculated to be 7. Therefore, the data do not provide sufficient evidence to reject the null hypothesis and to assert that there is a significant difference between the slopes of this valley.

The procedure for *large samples* (n greater than 25) is identical to that for small samples as far as, and including step 4—i.e. H_0, H_1, and the rejection level are stated, and the value of T is calculated in exactly the same way as that already described. The difference comes in deciding whether T is small enough to warrant the rejection of H_0. Appendix 7 does not give critical values of T for samples over 25. But it can be shown that, for samples above this size, the sampling distribution of T under H_0 approximates to the left-hand half (or tail) of a normal distribution with a mean equal to half the sum of all the ranks, which is $\frac{1}{4}n(n+1)$; and with a standard deviation of $\sqrt{\{n(n+1)(2n+1)/24\}}$. Therefore the probability (p) under H_0 of T being less than or equal to the value calculated can be found by:

(i) converting T into a z-score (difference of T from mean of sampling distribution *divided by* standard deviation of sampling distribution):

$$z = \frac{T - \frac{1}{4}n(n+1)}{\sqrt{\{n(n+1)(2n+1)/24\}}} \qquad \textit{Formula 6.10}$$

(ii) finding the value of p corresponding to the calculated value of z from Appendix 1 cols. B and C (giving one- and two-tailed probabilities associated with values of z in a normal distribution respectively).[5] A few moments' reflection will show that the *smaller* the value of T the further it is from the mean of the sampling distribution, and therefore the *greater* its z-score, and the *less* the likelihood (p) of its occurrence under H_0. If p is *less* than the rejection level (α) then H_0 can be rejected and H_1 adopted; if it is *more* than α H_0 cannot be rejected.

Example to illustrate the use of the Wilcoxon test for large samples
Table 6.14 contains commuting figures (columns A and B) for 30 rural districts with small populations in the East Midlands, taken from the 10 per cent sample census volumes of journey-to-work data for 1961 and 1966. They indicate an increase in the number of men commuting into rural districts, but there is the possibility that

TABLE 6.14

Number of working men enumerated as commuting into small rural districts in the East Midlands in 1961 and 1966

	$A(1961)$	$B(1966)$	$|A-B|$	Rank	$R_1(A>B)$	$R_2(A<B)$
Ashbourne	86	102	16	13		13
Bakewell	64	92	28	20·5		20·5
Clowne	58	59	1	1		1
Boston	73	98	25	18		18
E. Elloe	25	30	5	3·5		2·5
Spalding	68	88	20	16		16
E. Kesteven	85	123	38	26		26
S. Kesteven	59	80	21	17		17
W. Kesteven	115	110	5	3·5	3·5	
Caistor	36	84	48	28		28
Gainsborough	55	40	15	12	12	
Horncastle	69	65	4	2	2	
I. of Axholme	35	46	11	8		8
Louth	78	112	34	24		24
Spilsby	76	69	7	5	5	
Welton	97	128	31	22·5		22·5
Billesden	30	40	10	7		7
C. Donington	92	123	31	22·5		22·5
Lutterworth	98	155	57	29		29
M. Harborough	26	34	8	6		6
Brackley	25	87	62	30		30
Brixworth	50	67	17	14		14
Daventry	70	52	18	15	15	
Kettering	113	86	27	19	19	
Oundle	60	74	14	10·5		10·5
Towcester	66	101	35	25		25
Wellingborough	89	132	43	27		27
Ketton	57	45	12	9	9	
Oakham	43	72	28	20·5		20·5
Uppingham	8	22	14	10·5		10·5
	$\Sigma A = 1906$	$\Sigma B = 2315$		$T = \Sigma R_1 = 65·5$		

this increase is more apparent than real, because of sampling errors. The figures against each town would doubtless have been different if a different 10 per cent of households had been enumerated. One can, in fact, regard the actual counts as samples of 'populations' of possible 10 per cent counts for each census year, and use the Wilcoxon test to determine the probability that a difference as great as that between the 1961 and 1966 counts could have occurred if these hypothetical populations had been identical.

1. Null hypothesis (H_0): there is no difference between the number of men commuting into these rural districts in 1961 and 1966.

QUANTITATIVE TECHNIQUES IN GEOGRAPHY

QUANTITATIVE TECHNIQUES IN GEOGRAPHY

2. Alternative hypothesis (H_1): there is a difference . . .
3. Rejection level (α): 0·05
4. The test: the working is exactly the same as for small samples, and is set out in Table 6.14. It will be seen that T has been calculated as 65·5.
5. *The rejection or retention of H_0:*
   (i) Convert T into a z-score:

$$z = \frac{T - \frac{1}{4}n(n+1)}{\sqrt{\{n(n+1)(2n+1)/24\}}} = \frac{65·5 - \frac{1}{4} \times 30 \times 31}{\sqrt{(30 \times 31 \times 61/24)}} = \frac{-167}{49} = -3·4$$

(ii) Consult the z-table to find the probability of obtaining such a value of z in a normal distribution. As H_1 was non-directional a two-tailed probability is required, and the z-table (Appendix 1 col. C) gives a p value of 0·001 corresponding to a z-score of ±3·4. This is well below the rejection level of 0·05, so that H_0 can be rejected and H_1 adopted.

We have found, therefore, that the difference in the sample commuting figures in Table 6.14 almost certainly represents a real difference in the total numbers from which the samples were taken—i.e. that more men commuted into small rural districts in the East Midlands in 1966 than in 1961. Whether this is merely a manifestation of the increasing mobility of the working population, or of the dispersal of industry into rural areas, or whether it owes something to both these factors, could only be determined by further investigation. Statistical methods raise as many questions as they answer. It should also be pointed out that small East Midland rural districts are not necessarily representative of rural areas of varying population in other parts of the country. The rural districts listed in the table do not represent any broader category; one is only justified in making inferences about the whole population of male commuters in those particular areas.

Summary:
1. Formulate H_0 and H_1 and decide on rejection level.
2. Set out data as in Table 6.13. (Calculate differences between pairs of A and B; rank differences irrespective of sign, giving rank 1 to *smallest* difference; put ranks into two columns, headed $A > B$ and $B > A$; sum the columns to give ΣR_1 and ΣR_2.)
3. Whichever of ΣR_1 and ΣR_2 is smaller, call T.
4. (a) For *small* samples (up to 25 pairs) obtain significance of T from Appendix 7.
   (b) For *large* samples express T as a z-score (Formula 6.10) and obtain significance of z from Appendix 1 cols. B or C.

Consolidation
7. Take a 50 per cent systematic sample of rural districts from Table 6.14, and apply the Wilcoxon test to ascertain whether this sample is sufficient to yield a significant result.
8. Refer back to Table 6.7 (p.193), and use the Wilcoxon test to determine whether rates of change of population in large towns in Britain was significantly different during the periods 1951–61 and 1961–71.

6.8. THE ANALYSIS OF VARIANCE

We have seen above several methods of testing whether the means of two samples differ to such an extent that there is a high probability that they are drawn from different populations. It is often desired, however, to determine whether *more* than two (k) independent samples are derived from the same population. For example, in areas of highland Britain many scree slopes have accumulated, frequently in previously glaciated valleys. The scree generally consists of shattered rock fragments falling from a free face above and accumulating at a relatively steep angle at the base of the valley side. The cause of fragmentation is common to all, that is, prising away from bedrock by the freeze-thaw action of water and frost. But the source of the fragments varies, depending on the nature of the rock from which they come. The question is whether the angle at which the slope of the scree develops is related to the type of rock from which it is formed, rather than to the process of shattering common to all. We have a situation where one criterion (the angle at which the scree is resting) differs between a number of different samples, each of which represents a different kind of rock. The problem is whether the differences are such that we may infer they result merely from chance variations, or whether there is a high probability they are so great that the samples represent different populations. That is, screes developing from a variety of rock types form significantly different slope angles.

One method by which it is possible to test a number of samples is known as the analysis of variance. The most powerful of these techniques is the parametric 'F' test. However, not only must all the assumptions related to a parametric test be met, but the calculation is rather complex and time-consuming. (For these reasons it is not considered here.) Fortunately, there is a distribution-free test that can be used—the Kruskal-Wallis analysis of variance by ranks, which has the advantages of being distribution-free, quick, and simple to calculate, and suitable for use with a few small samples, as well as with many large ones. The Power-efficiency is 95·5 per cent.

A 'few small samples' is defined as a situation in which there are not more than three samples (i.e. $k = 3$), and not more than five values in each. Under these circumstances, to establish an exact probability that the three samples may be considered to be derived from the same population, we may compute a statistic, H, from the formula:

$$H = \frac{12}{N(N + 1)} \sum_{}^{k} \frac{R^2}{n} - 3(N + 1) \qquad \textit{Formula 6.11}$$

where R is the sum of the ranks in each sample, n is the number of values in each sample, N is the total number of values, k is the number of samples.

It is clear that the value of H for any particular set of samples (where N and n are given) will depend on ΣR^2. Whereas ΣR over all the samples will be the same however values are distributed among the columns, ΣR^2 will be higher the more the column values of R vary. For instance, $10 + 2 = 6 + 6$, but $10^2 + 2^2$ is much greater than $6^2 + 6^2$.

Consider the following data representing the angles of unvegetated scree slopes developed from three different rock types.

TABLE 6.15

	Limestone	Rank	Schist	Rank	Granite	Rank
Angle of	17	1	21	5	30	12
slope	20	4	18	2	26	10
	19	3	25	9	22	6
	24	8	28	11		
	23	7				
Sum of ranks	(R)	23		27		28

Note: Ranking may start from either the smallest or largest value.

1. Our null hypothesis, H_0, will be that there is no difference in terms of slope angle between the screes developed from the three different types of rock. (In other words the observed differences in angle between the three samples have a high probability of being due to chance variations.)

2. H_1 states alternatively that the angular variations between samples are so great there is a high probability that they are drawn from different populations: i.e. the differences are so great they are unlikely to be the result of 'chance'.

3. Our level of rejection is decided at $\alpha = 0.05$.

4. To test H_0 we calculate the statistic H using the formula for the analysis of variance, Formula 6.11.

$$H = \frac{12}{N(N+1)} \sum_{n}^{k} \frac{R^2}{n} - 3(N+1)$$

$$= \frac{12}{12(12+1)} \left(\frac{23^2}{5} + \frac{27^2}{4} + \frac{28^2}{3} \right) - 3(12+1)$$

$$= \underline{3 \cdot 26} \ldots$$

5. Reference to Appendix 8 shows that with the three sample sizes $n_1 = 5$, $n_2 = 4$, $n_3 = 3$, to be significant at the 0·05 level H must be equal to or *greater* than 5·6308. On the basis of the sample data in this example it is therefore not possible to reject H_0.

The same technique can be used for large samples, with $k > 3$, and with $n > 5$ (that is when the number of samples is three or more, and the number of values in any one is more than 5). Under these circumstances H is distributed as χ^2, with $df = k - 1$, and Appendix 4 may be used to determine the significance of the result.

Such a situation might arise in the analysis of beach deposits by size in order to establish whether these had a common origin. For example, six beaches are selected and a random sample of beach deposits taken from each, approximately half-way between mean high and low water. The samples are dried and equal quantities passed separately through mechanically shaken sieves, the proportion by weight in each ϕ unit being recorded. (ϕ units are simply a convenient way of classifying deposits by size. They are derived from Wentworth's millimetre system and based on a logarithmic scale. A full description of this method may be found in King (1966), p.274.) Let us suppose that the figures obtained are those set out in Table 6.16, and that we allot a rank to every value either the smallest or the largest value occupying the first rank. Any tied ranks will take the mean rank value.

Each column in Table 6.16 represents the sample taken from one beach. The left-hand figure is the weight in grammes for that sample of particle size shown in ϕ units for the row. The right-hand figure is the rank order of that weight through all six samples. Fortunately it is not necessary for each sample to have the same *number* of values, as in this kind of analysis it is not uncommon for some samples to have no fraction representing certain sizes, especially at either end of the range. Notice also that if the same *weight* of material had been used in each sample all columns possessing the same number of rows would have the same arithmetic mean, irrespective of how the

TABLE 6.16
Weight and grain size of sand samples from six beaches

Beach	A		B		C		D		E		F	
ϕ Units	Wt.	R	Wt.	R	Wt.	R	Wt.	R	Wt.	R	Wt.	R
−2 to −1	83	17·5	172	36	96	23	185	38	0	4	177	37
−1 to 0	93	22	139	32	0	4	79	16	61	11	203	39
0 to 1	0	4	150	33	0	4	132	31	83	17·5	88	20
1 to 2	104	25	155	34	55	9·5	123	30	52	8	85	19
2 to 3	70	14	107	26	92	21	209	40·5	209	40·5	114	28
3 to 4	111	27	99	24	68	13	119	29	0	4	221	42
4 to 5	72	15	168	35	65	12	0	4	55	9·5	0	4
ΣR_k		124·5		220		86·5		188·5		94·5		189

values are distributed within them. (In this case some fractions were lost and recorded weights are therefore unequal.) This does not matter because an analysis of variance depends upon the *variance* of the sample.

1. H_0 may now be formulated: that the variations in the amounts of particles of different sizes are due to 'chance', and that the samples are drawn from one population.

2. H_1 is that the samples are sufficiently different to warrant the conclusion that they come from more than one population.

3. The rejection level is fixed at 0·01.

4. The analysis of variance test is applied as before, using Formula 6.11.

$$H = \frac{12}{N(N+1)} \sum_{}^{k} \frac{R^2}{n} - 3(N+1)$$

$$= \frac{12}{42(42+1)} \left(\frac{124·5^2}{7} + \frac{220^2}{7} + \frac{86·5^2}{7} + \frac{188·5^2}{7} + \frac{94·5^2}{7} \right.$$

$$\left. + \frac{189^2}{7} \right) - 3(42+1)$$

$$= 14·87 \ldots$$

5. We have seen that with large samples the sampling distribution of H is similar to that of χ^2, with degrees of freedom (*df*) equal to $k - 1$. Reference to Appendix 4 shows with $df = 5$ that 14·87 is just not quite significant at the 0·01 level and therefore H_0 cannot be rejected. This does *not* mean, however, that the samples are from the same background population, for the critical value suggests that more than one population is involved, but with only 98 per cent, rather

than 99 per cent probability. In other words the test has indicated that the beach material has more than one origin, but has not quite reached the high level of probability we have declared essential to enable us to reject H_0. If research is to proceed further with the same rigour another, larger sample is required.

It will be observed that no allowance has been made for tied ranks. In fact, unless there are a large number of these (more than 25 per cent) the result is very little affected. What effect there is, is to make the H value slightly smaller, and therefore the test more rigorous. The correction for ties is simply to divide H by

$$1 - \{T/(N^3 - N)\}$$

where $T = \Sigma(t^3 - t)$, and t is the number of ranks that are tied in each group. In the example above, ranks 9 and 10 are tied and each represented by 9·5. T for this group is $T = 2^3 - 2 = 8 - 2 = 6$. The over-all sum of T therefore is

$$(7^3 - 7) + (2^3 - 2) + (2^3 - 2) = 336 + 6 + 6 = 348.$$

Substituting in the correction formula we have

$$1 - \{T/(N^3 - N)\} = 1 - 348/74046$$
$$= 0{\cdot}9953$$

H corrected for ties is therefore $14{\cdot}87/0{\cdot}9953 = 14{\cdot}94$.

It will be seen that the corrected value of H is almost unchanged in this case, but if there are many ties, and especially if these occur in long runs, this simple correction should be applied.

Summary
1. Formulate H_0 and H_1, and decide upon the rejection level.
2. Rank the data through all samples retaining the identity of the ranks of each sample.
3. Sum the rank values within each sample.
4. Apply Formula 6.11 to obtain the value of H.
5. If the sample is small with $k = 3$ and $n \leqslant 5$ an exact probability can be obtained from Appendix 8. If the sample is large with $k > 3$ and $n > 5$ the sampling distribution approximates to chi square with $df = k - 1$, and a significance level can be obtained from Appendix 4.

6.9. THE RAYLEIGH TEST FOR PREFERRED ORIENTATION
The problem of obtaining a preferred orientation was dealt with in Chapter 1 (p.13). The question now arises, as with the results obtained from any sample, of assessing its significance. Data are in

the form of frequencies distributed in class orientations of (in the example given) twenty degrees. The significance of the *distribution* of these frequencies can therefore be calculated by a simple Chi square test. But a better method is to test the preferred orientation itself through the use of a technique devised by Lord Rayleigh. The value required is termed $L\%$ where

$$L = \frac{R}{n} \times 100$$

and R is $\sqrt{\{(\Sigma f \sin 2\theta)^2 + (\Sigma f \cos 2\theta)^2\}}$, and n is the number of observations.

Substituting the values for the example given in Table 1.4 we have

$$L = \sqrt{\frac{(15 \cdot 481^2 + 27 \cdot 206^2)}{105}} \times 100$$
$$= 30\%$$

The significance of 30 per cent with $n = 105$ may be obtained from Appendix 12b. It will be seen to be at the 0·001 level. Had the sample contained only 50 observations a value of $L = 30\%$ would have been significant at the 0·01 level exactly.

Consolidation

9. The location of artifacts of an ancient culture were plotted on a map. Their distribution in terms of aspect and altitude was found to be as follows:

Height in metres	0–50	51–100	101–150	151–200	201–250	251–300	301–350	351–400
N	6	12	5	30	36	6	7	1
S	11	15	30	37	43	40	19	10
E	6	8	20	40	22	13	3	0
W	0	6	14	8	30	22	6	7

Aspect appears to have an effect upon the location of finds. Calculate whether this distribution is likely to be the result of 'chance' and the level of significance of your answer.

7

CORRELATION

CORRELATION is a word much used loosely in conversation. In writing of urban disturbances in the U.S.A. (*Guardian*, 17. Sept. 1971) William Greider reported an expert as saying: 'The more unemployment, the fewer the riots. There is almost a direct correlation over the past three years.' What the expert meant was that when unemployment was high there was less rioting (he does not imply there were *no* riots), and that in time of fuller employment there was more. As one variable (unemployment) increased, the other variable (riots) tended to decrease. In other words there appeared to be a relationship between the two variables, but it was inverse, not direct.

Or we may read that 'there is a strong correlation between cereal yields and the amount of June rainfall.' This implies that there is *an association* between the figures for rainfall and the weight of cereal yield, one series of values varying with the other. But what is meant by 'an association'? The answer is that the fluctuations in the values for each variable were sufficiently regularly (though not invariably) matched to make it unlikely that this was a chance association. Correlation in statistics is a method whereby a coefficient is calculated to describe the degree of association between two sets of paired values, and then tested to determine the probability that the association might be due to chance variations.

It must be clearly understood from the beginning, however, that because changes in the values of paired variates can be shown to be significant, that is, have less than (say) a 5 per cent probability of being random, this does not mean that one variable is *causing* the fluctuations in the other. No *causal* link can be deduced from a correlation coefficient *alone*. Cause and effect can only be decided through other evidence and the judgement of the observer.

In the last quarter of the nineteenth century Sir Francis Galton defined correlation and used r as the symbol for the correlation coefficient. The reason for his choice is because of the connection between correlation and regression (p.247 below). Correlation may be termed perfect and positive if one variate increases in precisely the same proportion as the other. It is termed perfect and negative if one variate decreases in the same proportion that the other increases. Take, for example, two variables, a and b.

a	1	2	3	4	5
b	3	6	9	12	15

In this situation we have perfect positive correlation. If the values were distributed thus:

a	1	2	3	4	5
b	15	12	9	6	3

we should have perfect negative correlation. Figure 7.1 shows the two situations plotted graphically. Correlation coefficients have been devised so that perfect positive correlation has a coefficient of $+1\cdot0$, and perfect negative correlation of $-1\cdot0$. If the association of the two variables is random the coefficient will be near 0. In the example of unemployment and rioting a significant (though not perfect) negative correlation is implied.

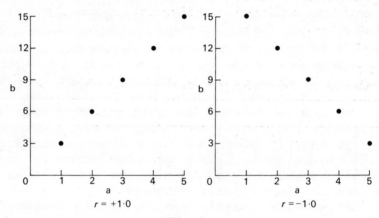

$r = +1\cdot0$ $r = -1\cdot0$

Figure 7.1.

7.2. THE PRODUCT MOMENT CORRELATION COEFFICIENT r

This is the most powerful test of correlation, but it is a parametric test and all the necessary conditions must be satisfied before it can be used. Fortunately, other methods of calculating correlation are available, but because the widely used Spearman Rank Coefficient (7.3) is derived from it, and as there are occasions when it can properly be used in geographical problems, calculation of the product moment coefficient is briefly outlined here.

So far in this book we have mostly used data which have the initial appearance of being distributed in some particular way. Let us on this occasion use data, by way of example, which we know to be random, by taking twelve pairs of random numbers, and calling them a and b (Table 7.1a).

TABLE 7.1a

a	80	61	23	94	87	37	64	22	23	42	17	39
b	30	29	33	21	61	56	86	69	22	38	18	45

The first step is to find the amount by which each variate differs from the mean of the variable and then to sum the product of the difference. We also require to know the standard deviation of a and b. The product moment correlation coefficient is calculated from the formula:

$$r = \frac{1/n \, \Sigma (a - \bar{a})(b - \bar{b})}{\sigma_a \cdot \sigma_b} \qquad \textit{Formula 7.1}$$

What we are doing here is to find the sum of the product of the total variations from $\bar{a}$ and $\bar{b}$ and dividing by the number of pairs, n. This is called the 'co-variance'. We reduce the co-variance to a ratio (the correlation coefficient) by dividing it by the product of the two standard deviations. Table 7.1b shows the calculation using the values given above.

TABLE 7.1b
Data tabulated for calculating product moment correlation coefficient

a	b	$(a - \bar{a})$	$(b - \bar{b})$	+	$(a - \bar{a})(b - \bar{b})$	−
80	30	30·92	−12·33			−381·24
61	29	11·92	−13·33			−158·89
23	33	−26·08	−9·33	243·33		
94	21	44·92	−21·33			−958·14
87	61	37·92	18·67	707·97		
37	56	−12·08	13·67			−165·13
64	86	14·92	43·67	651·56		
22	69	−27·08	26·67			−722·22
23	22	−26·08	−20·33	530·21		
42	38	−7·08	−4·33	30·66		
17	18	−32·08	−24·33	780·51		
39	45	−10·08	2·67			−26·91

$\bar{a}$ 49·08 $\bar{b}$ 42·33   2944·24   −2412·53
σ_a 26·14 σ_b 20·53   = 531·71

$$r = \frac{1/n \, \Sigma (a - \bar{a})(b - \bar{b})}{\sigma_a \cdot \sigma_b}$$

$$= \frac{531·71/12}{26·14 \times 20·53}$$

$$= \underline{0·08} \ldots$$

The correlation of random numbers will rarely work out to exactly 0, because '*a*' values and '*b*' values are *randomly* distributed. The coefficient, however, should always approach 0. In this case $r = 0.08$ is sufficiently close to zero to indicate that the numbers are randomly paired. Figure 7.2 is plotted from Table 7.1a, and shows graphically the appearance of a random distribution.

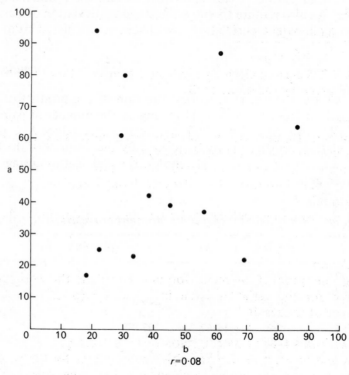

Figure 7.2.

A correlation coefficient is a statistic which *describes* the degree of association between two sets of paired values, and is used as such here. Usually the coefficient is subject to a test for significance like any other technique in inferential statistics. In this case, however, because we have chosen to correlate a sample of numbers known to be random, the coefficient is likely to be near zero. A population of random numbers tends towards an even distribution and therefore a test for significance is not valid. The method of testing the significance of *r* where this may properly be applied (i.e. when a population distribution approximating to normal may be assumed) is the same as that for Spearman's rank correlation and is given in 7.3 (p.225).

The use of Formula 7.1 to obtain r is rather laborious and inconvenient unless some form of calculating machine is available to do the arithmetic. Fortunately, as with the standard deviation, a more suitable formula for machine computation may be found, and

$$r = \frac{1/n\,\Sigma(a-\bar{a})(b-\bar{b})}{\sigma_a \cdot \sigma_b}$$

becomes

$$r = \frac{\Sigma(a \cdot b)/n - \bar{a} \cdot \bar{b}}{\sigma_a \cdot \sigma_b} \qquad\qquad \textit{Formula 7.2}$$

Formula 7.2 is derived algebraically from Formula 7.1 and will therefore yield precisely the same result.

Summary
1. Formulate H_0 and H_1 (in practice r will always be tested for significance), and decide upon the rejection level.
2. Calculate the mean and standard deviation of each variable.
3. Substitute in Formula 7.1, or if a calculating machine is available, in Formula 7.2.
4. Test the resulting coefficient for significance using Formula 7.4 (p.226).

Consolidation
1. Using the product moment formula calculate the correlation coefficient for the variables given in Table 7.2 (p.225). Test the significance of the result.

7.3. SPEARMAN'S RANK CORRELATION COEFFICIENT r_s

If it is found in practice that the rigid requirements for the product moment correlation cannot be assumed, and that some other method is necessary, we have in Spearman's rank correlation coefficient a well-tried alternative, having the characteristics of a distribution-free test. It gives a coefficient closely approximating to the product moment, having a power-efficiency of about 91 per cent compared with r, and has the added advantage of being quick and easy to calculate. It is called a 'rank' correlation because rank order is used to determine the association between the two sets of values, and not the actual values themselves. This method can thus be used if data on an interval scale are allotted ranks, and also when only ordinal measurement is possible, or when the margins of error are sufficiently great to make the use of an interval scale unrealistic. Situations of this kind occur sufficiently often in geographical work to make the technique very valuable.

The calculation of r_s is simple enough. First the variates are placed in rank order. Rank 1 is normally allotted to the highest value of each variable, Rank 2 to the next, and so on. The differences in rank of paired values are squared and summed. The correlation coefficient may then be found by the formula:

$$r_s = 1 - \frac{6\Sigma d^2}{n^3 - n} \qquad\qquad \textit{Formula 7.3}$$

where d is the difference in rank of each pair of values, and n is the number of pairs.

Figures for a worked example of r_s are given in Table 7.2. The correlation is between two variables of some importance—percentage population growth, and per capita percentage growth of Gross National Product. Medium-sized countries were chosen—defined as those with populations between twelve and ninety millions—in an effort to compare like with like as far as possible. Very large and very small countries were excluded, because in the calculation of r_s the values for each country have equal weight, and, were they included, the effect on the ultimate correlation coefficient of the figures for India with a population of 537 million would be in the same proportion as those for Trinidad and Tobago with a population of only 1 million (both mid-year estimates for 1969). The seven countries of Western Europe with the largest populations were chosen, and seven others randomly selected from those countries in the rest of the world whose growth rate per capita could be reasonably accurately estimated.

1. H_0 is that no relationship exists between population growth and the per capita growth of GNP in the countries selected.

2. H_1 is that there is a negative correlation between these variables.

3. We decide we must be able to reject H_0 with not less than 95 per cent confidence, therefore $\alpha = 0.05$.

4. The countries are tabulated in order of population size for convenience only—this does not affect the calculation. Percentage population growth (X) and per capita percentage growth of GNP (Y) is recorded for each. The X's are then allotted a rank, and similarly the Y's. Tied ranks are dealt with by the method given in Section 1.8a, p.42.

The difference in rank between each paired value is then recorded under column d, and the difference squared. The result of applying Formula 7.3 gives r_s as -0.55, revealing a negative correlation between the two variables. This indicates that as one tends to increase the other tends to decrease. In our example as percentage population

TABLE 7.2
Spearman's rank correlation coefficient r_s

Country	Population (millions)	% Growth rates popn.		% Growth rates per cap. GNP		d	d^2
		X	Rank	Y	Rank		
Brazil	88	3	2	1·6	10	8	64
Nigeria	62	2·4	4	−0·3	12·5	8·5	72·25
W..Germany	60	1·0	11	3·4	5·5	5·5	30·25
U.K.	55	0·7	14	2·0	8	6	36
Italy	52	0·8	13	4·0	3	10	100
France	49	1·1	9·5	3·7	4	5·5	30·25
Mexico	47	3·5	1	3·4	5·5	4·5	20·25
Spain	32	0·9	12	6·5	1	11	121
U.A.R.	31	2·5	3	1·6	10	7	49
Burma	26	2·1	6	1·6	10	4	16
Yugoslavia	20	1·1	9·5	4·2	2	7·5	56·25
Afghanistan	16	2·0	7	−0·3	12·5	5·5	30·25
Netherlands	12	1·3	8	3·0	7	1	1
Algeria	12	2·3	5	−3·5	14	9	81

$$r_s = 1 - \frac{6\Sigma d^2}{n^3 - n}$$

$$\Sigma d^2 = 707·5$$

$$= 1 - \frac{6 \times 707·5}{14^3 - 14}$$

$$= -0·55 \ldots$$

Figures were obtained from the *World Bank Atlas* (1970).

growth *increases* so percentage growth of per capita GNP *decreases*.
A glance at the figures will show that high percentage population
growth, as one would expect, is mainly associated with developing
countries. What the correlation coefficient reveals in economic, and
social, and political terms is outside the scope of this book. The
reader is left to consider this himself.

5. But is r_s statistically significant? Not all the developing
countries concerned have an adverse population/GNP growth rate.
For Mexico both are nearly the same. What then is the probability
that $r_s = -0·55$ is the result of a *chance* association? For samples of
between 4 and 30 pairs significance may be approximately determined from Appendix 9. The table shows that with $n = 14$ a critical
value of $r_s = 0·46$ is required for the 0·05 level of significance, and of
0·65 for the 0·01 level. We are therefore able to reject H_0 and to state
with 95 per cent confidence that there is a significant negative

correlation between the growth of population and the growth of per capita GNP, for those countries included in our study. Or, to put it another way, our result is significant at the 0·05 level, and therefore there are only 5 chances in 100 that a coefficient as high as this would occur from randomly paired data.

The significance of the product moment and Spearman's rank correlation coefficient may also be tested for a sample of any size having not less than 10 paired values by the use of the 't' table (Appendix 3). Note that correlation tests are normally one-tailed. Probabilities in Appendix 3 should therefore be halved.) The value of t is found by the formula:

$$t = r_s \,(\text{or } r) \bigg/\!\!\sqrt{\left(\frac{n-2}{1-r_s^2 \;\;(\text{or } r^2)}\right)} \qquad \textit{Formula 7.4}$$

Applying the formula to the coefficient calculated above with $r_s = 0·55$, and $n = 14$, we have:

$$t = 0·55\sqrt{\{(14-2)/(1-0·55^2)\}}$$
$$= \underline{2·28 \ldots}$$

Degrees of freedom in the case of a correlation coefficient are the number of pairs less two ($df = n - 2$). Reference to Appendix 3 shows a level of significance for this value of t between ·025 and 0·01 (one-tailed), the same as that previously obtained from Appendix 9.

Figure 7.3 shows the rank values for growth of population and per

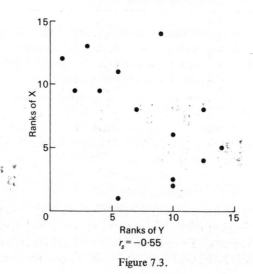

Figure 7.3.

capita GNP plotted graphically. This significant negative correlation may be compared with the random values shown in Figure 7.2.

Tied ranks affect r_s little unless there are a large number. There is a correction available for ties should this be required. (Space precludes its being given here, but a very lucid explanation may be found in Siegal (1956), p.206.) However, the reader is warned that the uncorrected coefficient becomes *less* rigorous as the number of tied ranks increases. If there are a number of these or if they occur in long 'runs', a safer correlation coefficient to use is Kendall's tau, described in 7.4 below.

Summary
1. Formulate H_0 and H_1, and decide upon the rejection level.
2. Allot ranks to each of the variates within variable X. Similarly allot ranks to each of the variates in variable Y. Set out the data in the form shown in Table 7.2.
3. Establish the difference in rank values between each pair of variates.
4. Square the differences found in (3) above.
5. Apply Formula 7.3.
6. Test the significance of the result using Appendix 9, or Formula 7.4.

Consolidation

	Annual rate of population increase	Daily calories per person	Grammes of protein per person
Argentina	1·5	3170	103
Bolivia	2·6	2060	52
Brazil	3·0	2700	67
Chile	2·4	2720	78
Colombia	3·2	2280	53
Ecuador	3·4	1850	47
Guyana	3·1	2290	55
Paraguay	3·3	2520	63
Peru	3·1	2300	55
Surinam	3·5	2510	62
Uruguay	1·2	3020	106
Venezuela	3·5	2490	60

2. In many areas of the world it seems that the very countries which have the most inadequate diet are those in which population is increasing most rapidly. The figures above are taken from the *United Nations Statistical Yearbook* for 1970 and show the annual rate of

population increase, and the daily per capita intake of calories and protein for countries in South America. Can it be said that there is an association between food consumed and population increase which is statistically significant? Is this true for calories as well as for protein?

7.4. KENDALL'S CORRELATION COEFFICIENT τ (TAU)

So far we have considered product moment correlation and Spearman's method derived from it. In Kendall's rank correlation the underlying basis is different. This is illustrated by the fact that in Spearman differences in rank are squared ($6\Sigma d^2$), thereby increasing the weight of large differences; in Kendall they are not. The coefficient is therefore not directly comparable numerically with r_s. Nevertheless, when the significance of τ and r_s is tested the result will be similar, because both have equal power-efficiency (91 per cent compared with r) to reject H_0.

Kendall's τ is a distribution-free statistic which is simple to calculate, may be used with very small samples. provided measurement at least on the ordinal scale has been achieved. It has the special advantage that it may be used in the calculation of a partial correlation coefficient; a method of considerable use to a geographer (see 7.5).

Let us consider τ in relation to small samples and assume that we wish to correlate two hypothetical variables, X and Y. First the paired values of each variable are allotted ranks, thus:

$$X \quad 2 \quad 1 \quad 3 \quad 5 \quad 4 \quad 6$$
$$Y \quad 1 \quad 2 \quad 4 \quad 5 \quad 3 \quad 6$$

Then arrange the ranks of X in natural sequence, setting below each rank the paired rank of Y. (The test is equally valid if either variable is placed in natural sequence with the paired ranks of the other placed below it.)

$$X \quad 1 \quad 2 \quad 3 \quad 4 \quad 5 \quad 6$$
$$Y \quad 2 \quad 1 \quad 4 \quad 3 \quad 5 \quad 6$$

Inspect the ranks of Y beginning at the left, in this case with the rank of 2. Record the number of ranks to the right that are greater than 2, allotting each +1. These are ranks 4, 3, 5, 6, which contribute +4. Record the number of ranks to the right which are smaller than 2, allotting each −1. In this case there is only rank 1, which contributes −1. Add the two contributions for rank 2 together, i.e. (+4) + (−1) = +3. Adopt the same procedure for each rank of Y. Sum the totals for all ranks of Y to form an over-all total termed S. In our example:

Ranks of Y

2	$(+4) + (-1)$	$= + 3$
1	$+4$	$= + 4$
4	$(-1) + (+2)$	$= + 1$
3	$+2$	$= + 2$
5	$+1$	$= + 1$
	$S =$	$+11$

The maximum possible score obtainable in this way, i.e. when the ranks form a perfect sequence, is found by $\frac{1}{2}N(N-1)$, where N is the number of ranks, τ is a measure of the disorder of the ranks of one variable when the other is placed in perfect sequence, and is found from the formula:

$$\tau = \frac{S}{\frac{1}{2}N(N-1)} \qquad\qquad \textit{Formula 7.5}$$

where S is the sum contributed by the ranks of one variable in the way described above, and N is the number of paired ranks.

It will be seen that when both variables have ranks in perfect sequence, i.e. when correlation is perfect, both the denominator and the numerator of the fraction will be equal, yielding a correlation coefficient of $+1$ or -1 depending on whether S is positive or negative.

In our example

$$\tau = \frac{S}{\frac{1}{2}N(N-1)}$$

$$= \frac{11}{\frac{1}{2}(6)(6-1)}$$

$$= 0 \cdot 73 \ldots$$

We now have to determine the probability that a coefficient value as high as $0 \cdot 73$ is due to chance variations. This depends upon the value of S, and the size of the sample, and may be determined from Appendix 10 for sample sizes 4 to 10. In our example the probability that, in a sample of 6, S will be equal to or greater than 11 is $0 \cdot 028$, or less than 3 per cent.

It will be observed that for small samples the calculation of S alone is sufficient to determine the significance of the association between two variables. It is not necessary to find τ, although this may be desirable for comparison with other samples. For large samples (with $N > 10$), the sampling distribution of τ approximates to the normal distribution, and probability may be established by calculating a z-value from the formula:

$$z = \tau \left/ \sqrt{\left\{\frac{2(2N+5)}{9N(N-1)}\right\}} \right.$$

Formula 7.6

Let us calculate τ for the data already recorded in Table 7.2, but rewritten so that the ranks of percentage population growth (X) are in the correct order, with the paired ranks ofgrowth of per capita GNP (Y) in the row beneath.

TABLE 7.3

X	1	2	3	4	5	6	7	8	9·5	9·5	11	12	13	14
Y	5·5	10	10	12·5	14	10	12·5	7	2	4	5·5	1	3	8

Both variables have tied ranks, arrived at by the averaging method described in section 1.8a. In calculating S (from variable Y) the right-hand rank(s) of a tie are allotted 0 when that rank is being considered. For example, the left-hand rank of Y, 5·5, is paired with all the ranks of Y to the right and allotted +1 or −1 accordingly, except when it is paired with the second 5·5 which contributes 0.

It will be seen that X also has two ranks tied at 9·5, corresponding to Y ranks of 2 and 4. The position of the ranks of Y is dependent on the position of the paired rank of X, but when two ranks of X are tied we are not able to determine to which one of the tie should be allotted 2, and to which 4. The order of the Y ranks of course affects S. In Table 7.3 had $Y4$ been placed before $Y2$ the value of S would have been increased by 2.

The solution is that, when considering the first rank of Y appearing under a tied rank of X, the Y rank paired with the second tied rank of X is accorded 0. So that in Table 7.3 in calculating S when we focus on $Y2$ and consider $Y4$ we record 0. If we had placed $Y4$ before $Y2$ we should still record 0. This is permissible because we are really counting both situations, i.e. $(+1) + (-1) = 0$.

We are now in a position to calculate S.

$$\begin{aligned}
S &= (8-4) + (3-7) + (3-7) + (1-8) + (-9) + (1-7) \\
&\quad + (-7) + (1-5) + (3-1) + (2-2) + (1-2) + (+2) \\
&\quad + (+1) \\
&= -33
\end{aligned}$$

Substituting $S = -33$ and $N = 14$ in Formula 7.5:

$$\tau = \frac{S}{\frac{1}{2}N(N-1)}$$

$$= \frac{-33}{\frac{1}{2}(14)(14-1)}$$

$$= -0.36 \ldots$$

Using Formula 7.6 we obtain a z value for τ

$$z = \tau \left/ \sqrt{\left\{\frac{2(2N+5)}{9N(N-1)}\right\}}\right.$$

$$= -0.36 \left/ \sqrt{\left\{\frac{2[(2)(14)+5]}{(9)(14)(14-1)}\right\}}\right.$$

$$= -1.8 \ldots$$

Reference to Appendix 1 col. B shows that with $z = 1.8$ the probability that $\tau = -0.36$ has occurred through 'chance' is 0.036, or less than 4 per cent. We may therefore accept $\tau = -0.36$ at the 0.05 level to reject H_0, i.e. the null hypothesis we had previously formulated for r_s (p.224) for the same data. (But note that although the *level* of significance is the same for both tests, τ is more rigorous than r_s, because ties are present.)

Unless they are numerous or occur in long runs, ties have no great effect upon the value of τ. However, when there is doubt concerning their effect a simple correction may be applied. For each group of tied ranks the correction may be found by $\frac{1}{2}t(t-1)$, where 't' is the number of ties in the group. For example, Table 7.3 shows Y as having three ranks tied at 10. The correction for this group, therefore, is $\frac{1}{2} \times 3(3-1) = 3$. The correction factor (C) for each variable is the sum of the correction values of each group of tied ranks, and may be expressed as:

$$C = \frac{1}{2}\Sigma t(t-1)$$

where t is the number of ties in each group.

In the above example the correction for X will be

$$C_x = \frac{1}{2}\Sigma t(t-1) = \frac{1}{2}(2)(2-1) = 1$$

and for Y

$$C_Y = \frac{1}{2}\Sigma t(t-1) = \frac{1}{2}(2)(2-1) + \frac{1}{2}(3)(3-1) + \frac{1}{2}(2)(2-1) = 5$$

The value of τ corrected for ties is found from the slightly modified formula:

$$\tau = \frac{S}{\sqrt{\{\frac{1}{2}N(N-1)-C_X\}}\sqrt{\{\frac{1}{2}N(N-1)-C_Y\}}} \qquad \textit{Formula 7.7}$$

where C_X and C_Y represent the correction for ties of each variable. The corrected value of τ in our example therefore is:

$$\tau = \frac{-33}{\sqrt{\{\frac{1}{2}(14)(14-1)-1\}}\sqrt{\frac{1}{2}(14)(14-1)-5}}$$
$$= \underline{-0\cdot375}\dots$$

The z-value obtained using Formula 7.6 is:

$$z = \tau\left|\middle/\sqrt{\left\{\frac{2(2N+5)}{9N(N-1)}\right\}}\right.$$
$$= -0\cdot375\left|\middle/\sqrt{\left\{\frac{2[(2)(14)+5]}{(9)(14)(14-1)}\right\}}\right.$$
$$= \underline{-1\cdot87}\dots \qquad \text{(a probability of } 0\cdot031\text{)}.$$

It will be seen the correction for the number of ties involved alters the probability only very slightly, and that τ still remains more rigorous than the uncorrected r_s.

Summary
1. Allot ranks to each of the values in variable X, and in variable Y.
2. Arrange the ranks of the one (X) in natural sequence, placing the paired ranks of the other (Y) below.
3. Calculate S by inspecting each rank of Y in turn and allotting +1 for every rank to the right which is greater; and −1 for every rank which is less. A tied rank considered with its tie yields 0.
4. Calculate τ, using Formula 7.5. If there is a large number of ties use the correction factor and Formula 7.7.
5. For sample sizes of 10 and below determine the significance of τ from Appendix 10. For samples larger than 10 calculate a z value using Formula 7.6.

Consolidation
3. As part of an investigation into the factors causing deposition by running water a natural stream was studied over a distance of 6 km. Upstream is a source of fine sandy material, which is readily identifiable, and which is transported and deposited by the stream in varying proportions along its course. The object is to try to establish whether the degree of sinuosity and length of wetted perimeter is connected with the amount of sand deposited. (The 'wetted perimeter' is the length of cross-section of the channel in contact with

water. Disregarding roughness it is a very approximate guide to the friction effect of the banks and bed on current velocity.)

The 6 km is divided into sections, each of 500 m, and a random sample of deposits is taken from the stream bed in each. These are sieved to determine the size of the sand fraction. As an indication of friction the apparent wetted perimeter is measured at each sampling point, and an index of sinuosity (similar to the detour index, p.63) is calculated for the length of stream 50 m above and below the sample. The (partly hypothetical) data collected are as follows:

Sample number	1	2	3	4	5	6	7	8	9	10	11	12
Sinuosity index	120	115	130	124	138	160	155	148	150	162	142	137
Percentage sand in sample	10	5	9	8	12	19	17	16	19	18	7	6
Length of wetted perimeter (in metres)	6·0	6·5	6·8	6·9	7·1	7·3	7·4	7·5	7·0	7·9	6·3	6·7

Using Kendall's rank correlation coefficient τ calculate the degree of association which can be said to exist between the proportion of sand in the samples and
   (a) the winding nature of the stream as measured by the sinuosity index,
   (b) the length of the wetted perimeter.
Comment on the significance of the result.

7.5. PARTIAL CORRELATION

One of the difficulties encountered by the geographer is that of having to assess the influence of each of a number of factors in a particular situation. Correlation is one way of deciding the relationship between *two* factors, but in many situations our interest lies with more than two. This may be exemplified from the field of agricultural geography.

Farmers decide upon the crops they grow or the animals they rear for a multiplicity of reasons. Most important are often factors such as a nearby market (like a large town, or food-processing industry). But tradition also plays a part, and some farmers tend to continue in the ways of the past. Physical conditions, such as climate, altitude, soils, and (especially in hilly areas) aspect, may also be important. Where other factors are marginal physical conditions may often determine the farmer's choice. The difficulty is to know the extent to which the correlation (and apparent causal relationship) between two variables is influenced by a third. Partial correlation is a method of dealing with three variables, in which the correlation coefficient of two of them can be measured unaffected by the influence of the third, which is controlled. As stated above one of the advantages Kendall's τ possesses is that it may be used in partial correlation.[1]

A situation in which this technique could usefully be employed arose during a study of dairy farming in an area in Yorkshire. It seemed plausible, in an area where physical conditions were relatively homogeneous, and the road system good, that the proportion of income derived from dairy produce might depend on two main factors, the size of the farm, and distance from the nearest large market. It was decided to calculate the correlation coefficients of the three variables under consideration—farm size, distance from market, and percentage income from dairying—and then to estimate the relationship between the percentage income from dairying and distance from the market, with the influence of farm size controlled. Similarly the relationship between farm size and percentage income from dairying would be worked out with the influence of distance from market controlled.

The formula for the partial correlation coefficient is:

$$\tau_{xy,z} = \frac{\tau_{xy} - (\tau_{xz} \times \tau_{yz})}{\sqrt{\{(1 - \tau_{xz}{}^2)(1 - \tau_{yz}{}^2)\}}} \qquad \textit{Formula 7.8}$$

where x, y, and z are different variables. The partial correlation coefficient $\tau_{xy,z}$ is calculated (i.e. the correlation coefficient between variables x and y, while the influence of the third variable, z, is controlled).

In the study x stands for farm size, y percentage income from dairying, and z distance from market. Correlation coefficients, calculated separately from data collected in the field, were as follows:

Distance from market—percentage income from dairying, $\tau_{yz} = 0.72$
Farm size—percentage income from dairying, $\tau_{xy} = 0.55$
Farm size—distance from market, $\tau_{xz} = 0.60$

The relationship between percentage income from dairying and distance from market, τ_{yz}, was tested first, with the influence of farm size controlled.

$$\tau_{yz,x} = \frac{\tau_{yz} - (\tau_{xz} \times \tau_{xy})}{\sqrt{\{(1 - \tau_{xz}{}^2)(1 - \tau_{xy}{}^2)\}}}$$

$$= \frac{0 \cdot 72 - (0 \cdot 60 \times 0 \cdot 55)}{\sqrt{\{(1 - 0 \cdot 60^2)(1 - 0 \cdot 55^2)\}}}$$

$$= \underline{0 \cdot 58 \ldots}$$

It will be observed that with the influence of the size of farm controlled τ_{yz} is reduced, showing that farm size has some effect on the proportion of dairying in the cases studied.

The next step was to test the relationship between farm size and the proportion of income from dairying τ_{xy} with the influence of distance from market controlled.

$$\tau_{xy,z} = \frac{\tau_{xy} - (\tau_{xz} \times \tau_{yz})}{\sqrt{\{(1 - \tau_{xz}^2)(1 - \tau_{yz}^2)\}}}$$

$$= \frac{0 \cdot 55 - (0 \cdot 60 \times 0 \cdot 72)}{\sqrt{\{(1 - 0 \cdot 60^2)(1 - 0 \cdot 72^2)\}}}$$

$$= \underline{0 \cdot 21 \ldots}$$

The effect of controlling the influence of distance reduces the value of τ_{xy} quite dramatically from $0 \cdot 55$ to $0 \cdot 21$, indicating that much of the apparent correlation between farm size and the proportion of income derived from dairying depends in part on the correlation of each with distance from market.

It should be noted that no test of significance has been devised for $\tau_{xy,z}$ because the sampling distribution has so far not been worked out. The coefficient therefore remains a descriptive statistic. For this reason no hypothesis was formulated before commencing the calculation.

This seems to be the appropriate place to emphasize once more that the *interpretation* of a statistical test remains a very important function of the researcher. Had distance not been taken into account in the above example, a quite erroneous conclusion might have been drawn. τ_{xy} was found to be significant at the $0 \cdot 01$ level. This coefficient revealed a mathematical relationship disclosing that the data correlated had a probability of less than 1 in 100 of being a 'chance' association; it concealed the fact that the relationship was not direct, but depending upon a third factor—distance. In this connection, although a significant correlation coefficient indicates some kind of causal relationship, it does not indicate the nature of the relationship, which may not be a matter of *direct* cause and effect.

Summary
1. Using Kendall's τ calculate the correlation coefficients between the three variables under consideration.

2. Apply Formula 7.8 to measure the extent to which the correlation coefficient between two of the variables is affected by their individual correlation with the third (controlled) variable.
3. Repeat this procedure with a different variable controlled.

Consolidation
4. With reference to the data given in Consolidation in 7.4 (p.233), and using the technique of partial correlation, establish the extent to which (a) sinuosity, and (b) the length of the wetted perimeter each contribute to the deposition of sand. Compare your results with those obtained from the original correlations.

7.6. THE POINT-BISERIAL CORRELATION

This correlation coefficient (derived from the product moment method) is particularly useful because of its great flexibility. Normally a correlation is determined by testing one series of values against an equal number of paired values to establish whether a relationship exists between the two series that is mathematically significant. The point-biserial technique enables a correlation coefficient to be calculated between a set of numerical values on the one hand and two categories on the other. These may, or *may not*, be expressed (or expressible) in figures. For example, it might be desired to compare the size of farms in a particular area with those in other areas, to determine whether significant relationships might emerge. The area chosen could be determined by altitude, rock type, administration boundary, or any criterion decided by the investigator. If the criterion chosen was limestone upland, then the correlation coefficient could be between 'limestone upland' and 'farm size', in which only the farm size is expressible in figures. This kind of technique is particularly useful in urban studies when frequently relationships are sought between an aspect of land use and some numerical variable (e.g. rateable value). The method is best demonstrated by an actual example, which is taken from the London area.

As part of a study of the service facilities in Kingston and Walton-on-Thames it was decided to investigate the difference in price between common grocery items sold in self-service stores (where the declared aim is generally cheapness achieved through scale economies), and the 'served' shop (where convenience, nearness to home, a delivery service, personal attention, and/or variety of goods held may be the attraction). The items to be used in the study were first chosen and then the prices charged by each shop in the area were recorded by type of shop (Table 7.4).

TABLE 7.4

Price in pence of a 'basket' of groceries in 'served' and 'self-service' shops

SS	S	S	S	SS	SS	S	SS	S	S
167	159	172	178	155	157	159	150	181	175
SS	SS	S	SS	S	S	SS	S	S	S
170	154	158	156	163	163	165	169	153	166

1. H_0 is that the prices charged for the same type of articles do not vary with the type of shop.

2. H_1 states that prices charged in served shops are significantly higher.

3. $\alpha = 0.05$ was decided as the required level of significance for rejection of H_0.

4. It will be observed there appears to be a relationship between shop type and price, in that the 'served' shops on the whole tend to be more expensive. This is what one would expect, but there are a number of exceptions (in this case related to the type of district in which the shops were located), and a statistical test is required to determine if the apparent relationship is significant.

First the data are dichotomized (i.e. divided into two) by the two categories of shop, as in Table 7.5.

The mean of each group of the dichotomized data is then calculated, together with the standard deviation of *all* the data, and the proportion of values in each group of *all* values. It is now possible to calculate the point-biserial correlation coefficient from the formula:

$$r = \frac{|M_p - M_q|}{S_x} \times \sqrt{(pq)} \qquad\qquad \textit{Formula 7.9}$$

where M_p is the mean of one group of values

M_q is the mean of the other group of values

$|M_p - M_q|$ is the difference between means irrespective of sign.

p and q are the proportions of cases in the two groups

S_x is the standard deviation of *all* the data.

Group p forms 12 out of the total of 20 observations. In this example therefore $p = 0.6$. Similarly with 8 observations $q = 0.4$.

$$r = \frac{|M_p - M_q|}{S_x} \times \sqrt{(pq)}$$

$$= \frac{|166.33 - 159.25|}{8.7} \times \sqrt{(0.6 \times 0.4)}$$

$$= \underline{0.40 \ldots}$$

TABLE 7.5
Dichotomized data for point-biserial correlation

Type of shop	Price of groceries X	Dichotomized data	
		p (served)	q (self-service)
SS	167		167
S	159	159	
S	172	172	
S	178	178	
SS	155		155
SS	157		157
S	159	159	
SS	150		150
S	181	181	
S	175	175	
SS	170		170
SS	154		154
S	158	158	
SS	156		156
S	163	163	
S	163	163	
SS	165		165
S	169	169	
S	153	153	
S	166	166	
		1996	1274
$S_x = 8.7$	Mean $p = 166.33$	Mean $q = 159.25$	

5. Reference to Appendix 9 shows that with $N = 20$ a coefficient of 0·377 is significant at the 0·05 level. (Alternatively Formula 7.4 may be used.) We are therefore able to reject the null hypothesis and to conclude with 95 per cent confidence that the average price of groceries sold in 'served' shops was higher than those obtained from self-service stores.

Finally, the point-biserial method is uniquely useful because it may be used in partial correlation (7.5). It is the only simple technique which makes it possible to include a 'criterion', not expressible in numerical terms, as one of the three variables involved. Formula 7.8 (which is employed in 7.5) may also be used with point-biserial coefficients, as it coincides with the formula used in the parametric product moment partial correlation.

Summary
1. Formula H_0 and H_1, and determine the level of rejection.
2. Calculate the standard deviation (S_x) of all values.

3. Separate the data into the two categories under consideration.
4. Determine the proportion of p and q of all the data.
5. Apply Formula 7.9.
6. Test the coefficient obtained for significance.

Consolidation
It is decided to investigate whether aspect could be said to be a factor affecting the steepness of slopes in areas of homogeneous rock type. A random sample is taken, and the aspect and mean angle of the principal slope element measured at 24 locations. The following data are recorded:

Aspect	N	N	S	S	N	S	N	S	N	N	S	N	S	S	N	N	S	S	N	S	S	N	S	N
Angle	7	11	9	5	17	8	13	13	21	9	10	9	12	7	14	9	4	6	18	9	12	22	4	14

5. (a) Calculate the correlation coefficient for slope angles on north- and south-facing slopes.
   (b) State H_0 and H_1.

6. At what level of significance is the result? What does this mean?

7. What statistical technique(s), other than correlation, might be used to determine whether aspect significantly affects slope angle?

7.7. THE CORRELATION MATRIX
When a number of variables are selected for correlation tests concerning a particular situation the results of the calculations are most conveniently set out in the form of a matrix. A good example of this method is the matrix (Table 7.6) devised by R. J. Robinson (1970) from ten variables concerning the economic situation in twenty-two countries in Latin America.

It will be seen from Table 7.6 that the variables chosen are socio/economic indicators for which information was available, and that the matrix is a quick visual method of detecting significant correlations. It is then possible to distinguish *groups* of variables where correlation is shown to be significant. For example, 3, 4, and 5 correlate at the 0·01 level, and indicate an association between these variables in providing a relatively high standard of living. While associated with the higher level of personal income, 4, are 6, exports (·01 level) and 9, industrial products (·05 level). For an interesting analysis and extension of these results the reader is referred to Robinson's paper of 1970.

TABLE 7.6

Correlation matrix of 10 variables for 22 Latin American countries

	1	2	3	4	5	6	7	8	9	10
1		−0·28	0·28	0·25	0·12	−0·32	−0·06	0·39	−0·28	−0·36
2			−0·13	−0·07	−0·05	0·01	0·18	0·31	0·10	0·68
3				0·91	0·63	0·32	−0·10	0·08	−0·28	−0·10
4					0·71	0·52	−0·01	0·01	−0·46	−0·16
5						0·25	−0·33	−0·10	−0·16	−0·31
6							0·18	−0·22	−0·27	0·19
7								0·01	−0·05	0·29
8									0·10	0·39
9										0·19
10										

Correlation at the 0·01 level of significance ☐

Correlation at the 0·05 level of significance ⌐ ⌐

Note: Only half the matrix is printed because the lower half is obviously an exact image of the upper half.

Variables:
1. Population (M) 1965.
2. Density of population per sq. km 1965.
3. Energy consumption per capita in kg coal equivalent 1965.
4. Income per capita 1964.
5. Calories per capita per day 1965.
6. Exports per capita 1965.
7. Population % growth per annum 1958–65
8. Agric. prodn. per capita, % growth per annum 1960–65.
9. Industrial prodn. per capita, % growth per annum 1958–65.
10. Cost of living increase % per annum 1960–64.

Source: *United Nations Statistical Year Book* (1966)

7.8. CORRELATION BONDS IN MORPHOLIGICAL ANALYSIS

Correlation techniques are also being increasingly used in the morphological analysis of landscape. It has been accepted by Chorley and Kennedy (1971) that if a large number of morphological variables correlate significantly, then this may indicate conditions of physical equilibrium.

They show evidence (Figures 7.4a and b) that where there is removal by streams from the base of slopes a high degree of significant correlation may be found of a number of variables. Whereas if the stream is away from the slope base the number of significant correlations is greatly reduced.

The implication of such an observation is that when stream and slope processes are operating in conjunction at present this promotes some kind of equilibrium

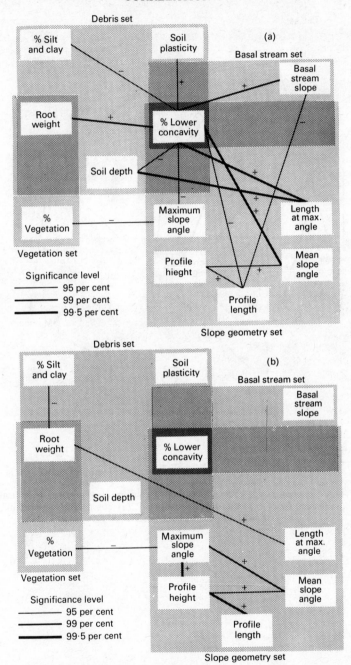

Figure 7.4. Correlation bonds (a) where debris is removed by steams; (b) where debris is not removed. From Chorley and Kennedy (1971).

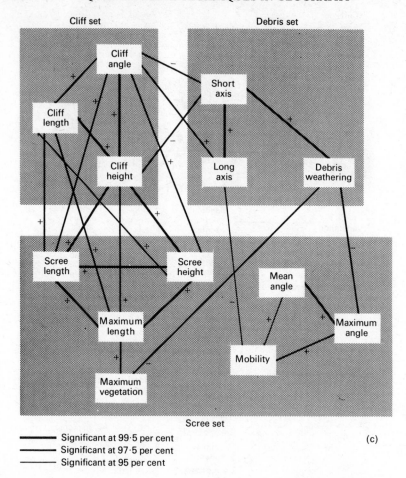

Figure 7.4. Correlation bonds between morphological elements which may be indicative of a local equilibrium state: (c) equilibrium conditions; (d) absence of equilibrium. From Chorley and Kennedy (1971).

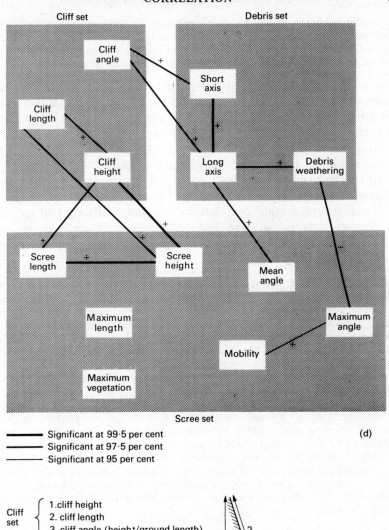

Cliff set   Debris set

Scree set

— Significant at 99·5 per cent
— Significant at 97·5 per cent
— Significant at 95 per cent

(d)

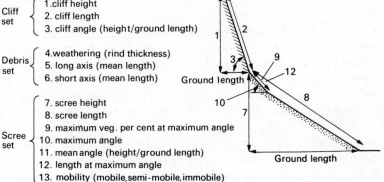

Cliff set
{
1. cliff height
2. cliff length
3. cliff angle (height/ground length)
}

Debris set
{
4. weathering (rind thickness)
5. long axis (mean length)
6. short axis (mean length)
}

Scree set
{
7. scree height
8. scree length
9. maximum veg. per cent at maximum angle
10. maximum angle
11. mean angle (height/ground length)
12. length at maximum angle
13. mobility (mobile, semi-mobile, immobile)
}

Ground length

Ground length

of operation within the system which is manifested by high correlations between all sorts of morphological parameters of stream and slope geometry, as well as debris and vegetation parameters. In short, the former case seems indicative of some kind of functional interaction, whereas the latter does not.

This is interesting, although probably to be expected if the concept of landscape equilibrium is acceptable. If there is a continual accumulation of debris over the lower part of the slope and on the valley floor, then the characteristics of the slope (e.g. in terms of angular measurements and length) will be constantly changing, and equilibrium cannot exist. Equilibrium implies that variables are 'adjusted' to each other, and, since the laws by which they relate and adjust are invariable, correlation should result. Figures 7.4c and d provide evidence of morphological correlations in a rather different situation. Here the contrast is between the development of scree slopes on Skiddaw Slates and Borrowdale Volcanics in the British Lake District. Chorley and Kennedy comment:

> . . . analysis of the cliff and basal scree systems . . . strongly indicate that some equilibrium conditions exist between the geometrical features of the cliffs, basal scree slopes and component debris of the Slates, which are absent from the similar features of the Volcanics. The natural extension of this argument is that forms are more related to *contemporary* (our italics) processes in the highly correlated systems, whereas the more poorly correlated are of largely relict systems on which the present processes are operating haphazardly and much less efficiently than previous processes. Of course, one has to exercise some care in drawing conclusions regarding the existence of a contemporary equilibrium state from observed statistical correlations in that some high correlations result from the operation of mutual constraints (i.e. such auto-correlation as that which tends to exist geometrically between slope length and slope height), and the remains of morphological adjustments to past processes would show similar high correlations simply because subsequent processes have been operating too weakly or for too short a period to change the former morphological relationships. This argument, however, applies much less forcibly to variables relating to soil and vegetation characteristics . . . [which adjust to changing conditions relatively quickly].

The search for correlation bonds to establish interlocking systems as methods of morphological analysis is a relatively new technique in geomorphology. It has been quoted at some length here because the measurements necessary in the field may be made with simple instruments that are readily available and easily portable. Even the soil analysis requires only the most rudimentary laboratory equipment. Calculation of the correlation coefficients using r_s or τ may be done quickly and easily. Investigations of this nature are thus within the reach of any interested student of geomorphology. Little work has so far been done, and there is great need for the accumulation and publication of factual information. The field is wide open for research at almost all levels. The interpretation of results, as noted above, must be approached much more cautiously. Confirmation of the explanations tentatively suggested must await the collection of many more field investigations, and it is in this respect that the geomorphology student can render an important service.

7.9. THE CHI SQUARE CONTINGENCY COEFFICIENT

The chi square contingency coefficient C is a kind of correlation coefficient which measures the degree of association between two or more sets of data. It may be used when data are measured on the nominal scale, provided that the conditions are met for the normal chi square test (p.171). The data are set out in the form of a contingency table, for example as in Table 6.3b, and the value of χ^2 calculated using Formula 6.2. The contingency coefficient may then be found by:

$$C = \sqrt{\frac{\chi^2}{\chi^2 + N}} \qquad\qquad Formula\ 7.10$$

where N is the total number of frequencies in the sample.

Using the data and χ^2 value found in the example on page 179, we find that

$$C = \sqrt{\frac{11 \cdot 03}{11 \cdot 03 + 110}}$$

$$= \sqrt{0 \cdot 09}$$

$$= \underline{0 \cdot 30}.$$

We now have a numerical index describing the degree of association between slope angles that have developed on Keuper Marl, and on Greensand. There is no test of significance for C, which may be regarded as having the same level of significance as the value of χ^2 from which it was calculated.

The chi square contingency coefficient possesses a number of advantages for the geographer. It may be used when measurement is only on the nominal scale. It enables a numerical index of the relationship between two *or more* variables to be established. (Although the example given here is only for two.) If thus allows comparisons to be made between different locations. It reduces data obtained from a number of variables to a single value that can be mapped. It may be used to summarize the relationship between maps, or mappable distributions. It is simple and relatively easy to calculate.

There are also disadvantages. C can only be used within the constraints imposed by the chi square test. Although the lower limit of C is near 0 (i.e. when the association between variables is random), the upper limit depends on the number of categories involved, and can never reach unity, even when the variables are perfectly correlated. For example, in a 2 $\times$ 2 contingency table the greatest value of C is 0·707, while in a 4 $\times$ 4 table it is 0·87. It is therefore essential when comparisons are to be made that the contingency tables are of the same size.

Summary
1. Cast the data in the form of a contingency table.
2. Calculate the value of χ^2 using Formula 6.2.
3. Calculate the value of C using Formula 7.10.
4. Establish the significance of χ^2 from Appendix 4.

8

REGRESSION

8.1. INTRODUCTION

A CORRELATION coefficient measures the degree of association between two sets of paired variates. A significance test may show whether that degree of association is likely to be more than a matter of chance. What neither the correlation coefficient nor a significance test tells us is the way in which the two sets of variates are related: they cannot be used to predict one set of variates from a knowledge of the other. Nor do they signal any anomalies in the relationship between individual pairs. If we wish to pursue these lines of inquiry we must turn to regression analysis and the study of residuals. It should, however, be stressed at the beginning that both are of limited value unless the variables in question are significantly correlated.

A regression line is simply a line of 'best fit' on a scattergram (Figure 8.2). It may be merely a summary expression of the relationship between two variables; or a means of highlighting individual deviations (called residuals) from this general relationship; or it may be used to interpolate or predict unknown values of one variable from known values of the other. The trend line plotted through a time series (Figure 3.8) is a special case of a regression line, where one of the variables is time (usually plotted at evenly spaced intervals) and the other variable a set of data of different magnitudes. More commonly, scattergrams and regression lines chart the relationship between two magnitude variables, neither of which is plotted at evenly spaced intervals. Similarly, the fluctuations about the trend line of a time series (Figure 3.9) constitute a special case of residuals or deviations from a regression line.

8.2. DEPENDENT AND INDEPENDENT VARIABLES

One thing that neither a correlation coefficient nor a regression line can show us is whether there is any direct causal relationship between variables. Common sense is a better guide. If there is a significant correlation between rainfall and crop yield, it is likely that the former has influenced the latter and highly unlikely that the latter has influenced the former. Crop yields depend on rainfall, but rainfall does not depend on crop yields. Therefore, when considering the relationship between these two variables, we may reasonably call crop yields the *dependent variable* and rainfall the *independent variable*. It is conventional when drawing graphs or scattergrams, to measure the independent variable along the horizontal (x) axis, and the dependent variable along the vertical (y) axis. Therefore, although a regression line may not indicate the nature of any causal relationship between variables, a knowledge or hunch about causal relations may affect the way the regression line is drawn.

On the other hand, there may be no direct causal relationship between two significantly correlated variables. It is doubtful whether the monthly sale of swim-suits in New Zealand is influenced by the sale of winter coats in Britain or vice versa—their correlation is due to their both being related to a third variable, the location of the over-head sun. In this case there is no obvious dependent or independent variable. If, however, we are going to use regression to predict values of one variable from known values of the other (strictly, interpolate or extrapolate—see section 8.4), then the convention is to treat the former (i.e. the predicted) as the dependent variable (measured along the y-axis) and the latter (i.e. the predictor) as the independent variable (measured along the x-axis).

8.3. SCATTERGRAMS, REGRESSION LINES, AND RESIDUALS
Consider the figures in Table 8.1:

TABLE 8.1

x	1	3	4	7	11
y	2	6	8	14	22

Let us first draw a scattergram showing the relationship between x and y (Figure 8.1).

The points are seen to be in a straight line, indicating perfect correlation ($r = +1$). There is no problem about drawing the 'best fit' line—it is clearly the line going through all the points. There are no residuals (anomalies) since every point lies on the line (a residual is shown on a scattergram by the amount of deviation of a point from the regression line). Residuals only occur if variables are not perfectly correlation—which is almost always in geographical problems.

Now consider the figures in Table 8.2:

TABLE 8.2

x	2	4	5	9	11
y	6	6	13	14	23

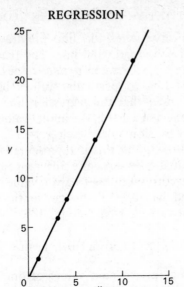

Figure 8.1. A scattergram of perfectly correlated paired variates

The scattergram in Figure 8.2 shows that the two sets of variates are not perfectly correlated, because the points representing them are not in a perfectly straight line. Nevertheless, there is not much difficulty in drawing a regression line by eye. One can then emphasize the deviations from the regression line by drawing residual lines (δ_1 to δ_5).

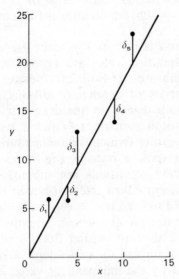

Figure 8.2. A scattergram with regression and residual lines

It may be wondered why the residual lines are drawn vertically rather than perpendicular to the regression line. The reason is that vertical lines show the difference between *actual* values of the dependent variable y corresponding to particular values of the independent variable x, and the values that the regression line might lead one to expect. For instance, when x is 11, the actual value of y in Figure 8.2 is 23; but on the regression line, when $x = 11$, $y = 20$. There is, therefore, a residual of 3, and this is represented by δ_5. In the same way, horizontal residual lines show the difference between actual and expected values of x corresponding to given values of y, but these are less commonly used because of the convention to represent the given or independent variable by x (see p.248).

8.4. INTERPOLATION, EXTRAPOLATION, PREDICTION, AND EXPLANATION

We have just referred to expected values of one variable, given the values of another with which it is correlated. Such expected values are obtained by *interpolation* (estimation of intermediate values from known data—e.g. values of y corresponding to values of x between 2 and 11 from the regression line in Figure 8.2) or by *extrapolation* (estimates of values beyond the limits of known data—values of y corresponding to values of x less than 2 or more than 11 in Figure 8.2). This form of statistical inference is also called *prediction*, but the word has a double meaning: the one used here, which implies a purely mechanical reading off a regression line, and the one described on p.106, which implies the exercise of judgement about future trends.

Statistical prediction is one of the main uses of regression. For instance, evapo-transpiration rates are known to correlate closely with wet-bulb temperatures. If both are measured on a number of occasions under a variety of weather conditions, the results plotted on a scattergram, and a regression line drawn, it will be possible to interpolate evapo-transpiration rates (which take time to measure) from wet-bulb temperature (which can be measured quickly). It may be necessary to extrapolate outside the range of points on the scattergram, if extreme conditions are subsequently encountered. Or again, if the mean rainfall of five stations in the upper part of a drainage basin is found to correlate closely with water-level at a point lower down the course of the trunk stream 24 hours later, a regression line can be used to predict flood levels after heavy rain.

In both the above examples, the dependent variables can be said to be partially *explained* by the independent variables, in that a knowledge of the latter (wet-bulb temperatures and rainfall) enables one to make a fairly accurate estimate of the former (evapo-transpiration rates and flood levels). But the word 'explained', like the word 'predict', has a specialized statistical meaning not to be confused with its more generally understood implications of cause and effect. It is widely used in connection with correlation and regression, and it must be understood to have the limited connotation of a statistical relationship enabling interpolations and extrapolations (predictions) to be made from one variable to another.

Some points regarding the reliability of predictions based on regression:

(i) The higher the correlation between two variables (the predictor and the predicted), the more accurately and confidently can predictions be made from one to the other. This is because a regression line will fit closely the array of points on the scattergram, so that residuals are small, if the correlation is high.

(ii) The best-fit regression line for an array of points is not necessarily a straight line. Two variables may be perfectly correlated in the sense that there is an entirely regular and predictable relationship between them, but the coefficient of linear correlation (r) will not be $+1$ or -1 if the relationship is not linear. A linear regression line may, indeed, be misleading, if it distracts from an important curvilinear relationship—a point that will be taken up again presently. Unfortunately, curvilinear regression lines are difficult to construct (except in a rough way by 'eyeballing'), and we shall only deal with linear regression. But curvilinear relationships can sometimes be straightened out by logarithmic transformations—see section 8.7.

(iii) Interpolation is generally safer than extrapolation, because there may be important thresholds determining the relationship between variables beyond the range of the given data— e.g. if density of water is given for temperatures at 10 degree intervals between 10°C and 90°C, interpolation of densities at other temperatures between 10° and 90° will be safe; extrapolation beyond that range may fail to take account of what happens to water at boiling and freezing points, as well as the critical temperature of 4°C.

(iv) Where correlation from a large number of measurements is perfect, interpolation can be made (but not necessarily extrapolation) with complete accuracy and confidence. One variable is said to 'explain' the other completely—the level of statistical explanation is 100 per cent, a rare occurrence in geography.

The way in which 'statistical explanation' may be quantified when correlation is less than perfect is best shown by referring back to Table 8.2. First, consider the five values of the dependent variable, y. These are 6, 6, 13, 14, and 23, and they have a mean ($\bar{y}$) of 12·4. The variability of y can be expressed in terms of the deviation of individual variates from the mean, as illustrated by the lengths of d_1, d_2, d_3, d_4, and d_5 in the dispersion diagram (Figure 8.3).

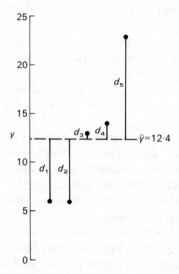

Figure 8.3. Dispersion diagram to show deviation of individual values of y about the the mean ($\bar{y}$)

The variability of y about $\bar{y}$ can be expressed by the Variance (Var), which is given by the formula

$$\text{Var} = \Sigma d^2/n \qquad\qquad\qquad \textit{Formula 8.1}$$

where d refers to the deviation of each variate from the mean, and n is the number of variates. (The variance is the square of the standard deviation.)

If we now refer back to Figure 8.2, we can express the variability of y about the regression line in the same way, substituting residuals for deviations about the mean:

$$\text{Variance about regression line} = \Sigma \delta^2/n \qquad \textit{Formula 8.2}$$

From a glance at Figures 8.2 and 8.3, it is clear that $\Sigma \delta^2/n$ is smaller than $\Sigma d^2/n$. The extent to which it is smaller is a measure of the degree to which the values of y have been statistically 'explained' (via the regression line) by x. This is generally expressed as a percentage of the variance:

Level of explanation of y by x (E)

$$= \frac{\{(\Sigma d^2/n) - (\Sigma \delta^2/n)\}}{\Sigma d^2/n} \times 100$$

$\therefore E = 100\{1 - (\Sigma \delta^2/\Sigma d^2)\}$ % of the Variance about the mean.

Formula 8.3

Applying the formula to Figures 8.2 and 8.3,

$\Sigma d^2 = (6\cdot4)^2 + (6\cdot4)^2 + (0\cdot6)^2 + (1\cdot6)^2 + (10\cdot6)^2 = 197$
$\Sigma \delta^2 = (2)^2 + (2)^2 + (3)^2 + (4)^2 + (1)^2 = 34$
$\therefore E = 100\{1 - (34/197)\} = 83\%.$

Therefore, in this hypothetical example, differences in the value of the independent variable x 'explain' 83 per cent of the variance in the dependent variable y. It should, however, be noted that this is only an approximate estimate, because, although values of d can be calculated accurately (by subtracting individual values of y from $\bar{y}$), values of δ have only been measured roughly from the residual lines on Figure 8.2; and these residual lines extend from a regression line which has been draw in by eye.

8.5. METHODS OF CONSTRUCTING REGRESSION LINES
The smaller the correlation between variables, the more difficult it becomes to draw by eye the 'best-fit' regression line through an array of points on a scattergram. If the line really is to be a best fit, a more precise construction technique is required. Three techniques will now be outlined, two of which are very simple and the third more sophisticated.

8.5a. Regression lines for ranked data
When paired data are ranked, the best-fit regression line is simply a diagonal, running from the origin if the data are positively correlated, and between the extremities of the axes if negatively. This is illustrated in Figure 8.4, which shows the relative importance of different types of shop in two towns of similar size but contrasting character. (Note that the scales appear to run backwards—this is because 21 is the lowest rank, and 1 the highest.)

The scattergram shows that the two sets of ranks are positively correlated ($r_s = 0\cdot82$), meaning that shops which are well represented in one town tend to be well represented in the other also. The regression line draws attention to the differences in relative importance

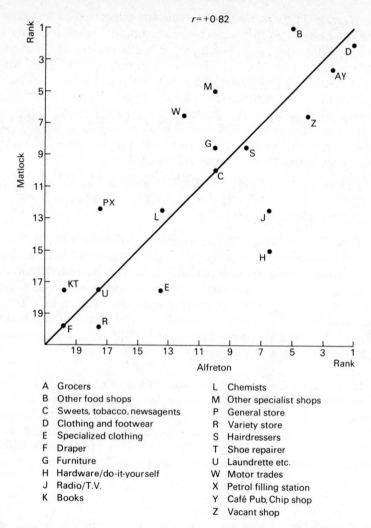

Figure 8.4. Scattergram of shop types ranked by frequency in Alfreton and Matlock

of shop types in the two towns: the further a point is from the diagonal, the greater the difference. Thus, hardware and do-it-yourself shops are relatively more numerous in Alfreton (rank 6·5) than in Matlock (15), while motor trade establishments are more numerous in Matlock (rank 6·5) than in Alfreton (12). This type of regression line is of limited value, but it does focus attention upon the differences between data sets.

8.5b. Regression by the method of semi-averages

Where data are on an interval scale, a method of reducing though not entirely eliminating, the element of guesswork in drawing a regression line, is by the use of semi-averages. Calculation is simple, and the result is more reliable than that achieved by relying solely on 'eye-balling'. This method has already been described in Chapter Three (p.96). But there the problem was simplified by the fact that the independent variable was represented by regularly spaced points in time. More usually this is not the case, and the method has to be modified accordingly.

Consider the data in Table 8.3, and plotted in Figure 8.5:

TABLE 8.3
A table of paired variates

x	1	3	4	7	10	12	15	$\bar{x} = 7 \cdot 4$
y	3	5	9	7	8	16	20	$\bar{y} = 9 \cdot 7$

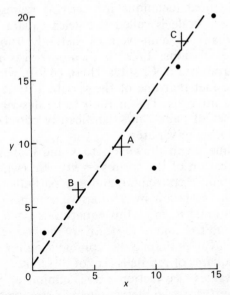

Figure 8.5. Scattergram of data in Table 8.3, with semi-averages regression line

To construct the regression line, three additional points are plotted: the mean (*A*), and the lower and upper semi-average (*B* and *C*).

1. The mean (*A*) is simply the point whose co-ordinates are equal to the means of *x* and *y* ($\bar{x} = 7 \cdot 4; \bar{y} = 9 \cdot 7$).

2. The lower semi-average (B) has co-ordinates which are determined as follows:
   (i) x co-ordinate: the mean of all values of x *below* $\bar{x}$—in this
   (ii) case $(1 + 3 + 4 + 7)/4 = 3.75$.
   (ii) y-co-ordinate: the mean of all values of y *below* $\bar{y}$—
   $(3 + 5 + 9 + 7 + 8)/5 = 6.4$.

3. The upper semi-average (C) has co-ordinates which are determined in a similar way to the lower semi-average: they are the
   means of all values of x and y above $\bar{x}$ and $\bar{y}$—viz. $(10 + 12 + 15)/3 = 12.3$, and $(16 + 20)/2 = 18$.

The three points A, B, C are almost in a straight line. The regression
line is now drawn by eye as a best fit for these three points. It is
clearly much easier to draw by eye a best-fit line through three
points which are themselves nearly in a straight line, than through a
large number of points with a much less linear scatter.

8.5c. *Regression by the least-squares method*
The most widely used technique for deriving regression lines from
data on an interval scale is called the 'least squares' method. It is
more rigorous than the semi-averages method although it involves
more lengthy calculations. Like the semi-averages method, it has
already been introduced in Chapter Three (p.99), where again it was
simplified by the fact that one of the variables, time, was measured
at evenly spaced intervals. The method to be described now can be
applied to any set of paired and significantly correlated data, irrespective of the spacing of variates.

Any straight-line graph drawn on to x and y axes, can be represented by an equation of the form $y = mx + c$ (where x and y are
variables, and m and c constants); and any particular straight line is
represented in the equation by a unique combination of m and c.
(Another conventional form of this equation is $y = a + bx$, in which
a and b correspond to c and m respectively in $y = mx + c$.) The least
squares method involves finding this combination for a line such that
the sum of the squares of residuals (i.e. of differences between actual
and 'predicted' values—see section 8.3) is minimized.[1] For any set of
paired data, there are two 'least squares' regression lines, which
minimize the sum of the squares of residuals for the x- and y-variables
respectively.

If values of y are to be 'predicted' from given values of x, the line
which minimizes y-residuals is used; if values of x are to be predicted
from given values of y, the line which minimizes x-residuals is used.
This is illustrated in Figure 8.6, in which AB is called the regression

line of y on x (i.e. the line which is used for predicting values of y for given values of x); and CD is called the regression line of x on y (i.e. the line which is used for predicting values of x for given values of y). AB is the line for which $\Sigma\delta_y^2$ is minimized; CD is the line for which $\Sigma\delta_x^2$ is minimized. Because of the convention of assigning the independent variable (or that for which values are given) to the x-axis, and the dependent variable (or that for which values are to be predicted) to the y-axis, the regression line of y on x (AB) is more commonly used. The equation of this line is given by the formula

$$y - \bar{y} = r \cdot \frac{s_y}{s_x}(x - \bar{x}) \qquad\qquad \textit{Formula 8.4}$$

where r is the produce moment correlation coefficient; $\bar{x}$ and $\bar{y}$ the means, and s_x and s_y the standard deviations of the given values of x and y.

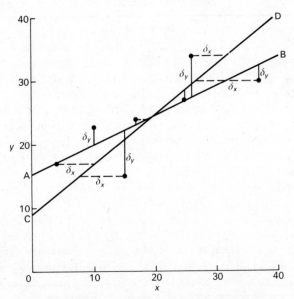

Figure 8.6. Scattergram with two least-square regression lines

This formula can be expressed in the form $y = mx + c$ as follows:

$$y = \left(r \cdot \frac{s_y}{s_x}\right)x + \left(\bar{y} - r \cdot \frac{s_y}{s_x}\bar{x}\right),$$

the two terms in brackets being the values of m and c. (It should be noted that the symbols in Formula 8.4 are those used for samples;

if the calculations are based on population data, then μ and σ are the appropriate symbols for the mean and standard deviation.)

Formula 8.4 gives the regression equation for y on x; the corresponding formula for the regression equation of x on y is

$$x - \bar{x} = r \cdot \frac{s_x}{s_y}(y - \bar{y}) \qquad\qquad \textit{Formula 8.5}$$

The method will now be illustrated with an example. Table 8.4 lists some employment and education statistics for a random sample of sixteen towns, taken from Moser and Scott (1961).

TABLE 8.4
Employment and education data (1951) for sixteen British towns

		x % employed in Industrial Orders I to XVI	y % left school before age 15
1	Acton	67	67
2	Blackburn	60	75
3	Burnley	60	75
4	Colchester	29	72
5	Doncaster	48	59
6	Finchley	25	45
7	Hastings	12	61
8	Ipswich	40	70
9	Luton	68	70
10	Newport	39	66
11	Plymouth	29	59
12	St Helens	62	82
13	Southgate	24	36
14	Swansea	40	66
15	Walsall	57	80
16	Willesden	53	62

Mean of $x\,(\bar{x}) = 44\cdot6$ Mean of $y\,(\bar{y}) = 65\cdot3$

Standard deviation   Standard deviation

$I = 0\cdot64$ of $x\,(s_x) = 16\cdot6$ of $y\,(s_y) = 11\cdot5$

The product moment correlation coefficient shows that there was a positive correlation of $0\cdot64$ in these towns in 1951 between the proportion employed in the primary and manufacturing sectors (Industrial Orders I to XVI) and the proportion of residents who left school before the age of 15. This is significant at the $0\cdot01$ level. The equations of the regression lines can be calculated as follows:

1. For y on x, using Formula 8.4:

$$y - 65\cdot3 = 0\cdot64 \times \frac{11\cdot5}{16\cdot6}(x - 44\cdot6)$$

which becomes $y = 0\cdot44x + 46$

2. For x on y, using Formula 8.5:

$$x - 44\cdot6 = 0\cdot64 \times \frac{16\cdot6}{11\cdot5}(y - 65\cdot3)$$

$$\therefore \qquad x = 0\cdot92y - 15\cdot5$$

These equations have been plotted as straight lines on the scatter-gram (Figure 8.7) by simply taking two values of the variable on the

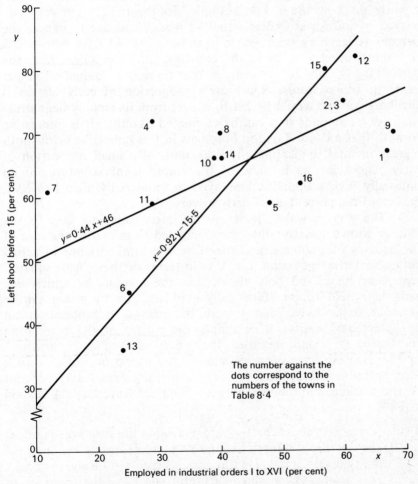

Figure 8.7. Least-squares regression lines for data in Table 8.4

right-hand side of each equation, calculating corresponding values of the other variable, plotting the resulting co-ordinates, and ruling lines through these points.

Compared with the method of semi-averages, this is a lengthly procedure, involving the calculation of standard deviations and the product moment correlation coefficient before the equations can be determined. But it has two advantages:

1. It is possible to predict a value of one variable from a given value of the other by using the equations directly: it is not necessary to draw lines or measure anything on a scattergram. Similarly, residuals can be calculated rather than measured. Results are consequently more accurate. For example, Southgate's 24 per cent employed in Industrial Orders I to XVI would lead one to expect the percentage leaving school before 15 to be: $y = 0 \cdot 44 \times 24 + 46 = 57\%$ (by substituting $x = 24$ in the equation for y on x). In fact, the percentage is only 36 per cent, so that there is a residual of 21 per cent. In other words, Southgate's proportion of early leavers is smaller than one would be led to expect from its employment structure, by an amount that can be calculated exactly. (It is interesting to note that all four London boroughs in this sample lie below both regression lines, indicating either an unusually small proportion of early school leavers in view of the employment structure; or an unusually high proportion employed in Industrial Orders I to XVI, in view of the percentage of early leavers.)

2. The way in which 'least squares' regression lines best fit the data is known exactly: they are calculated to minimize the sum of the squares of residuals, and therefore the total variance of actual data about the regression line. It is in terms of these 'least squares' regression lines, and only these, that the amount by which one variable is said to be 'statistically explained' by the other can be determined precisely. What is more, the amount of explanation can be determined easily: it is simply the square of the correlation coefficient (r^2), and is called the 'coefficient of determination'. In the case we have been examining, the proportion of the work force in Industrial Orders I to XVI statistically explains 41 per cent of the variance in the percentage of school leavers ($0 \cdot 64 \times 0 \cdot 64 = 0 \cdot 41$).

Two other points about least-square regression lines are worth noting:
   (i) The two lines intersect at the point representing the mean of the two sets of data (in this case, where $x = 44 \cdot 6$ and $y = 65 \cdot 3$). This is always so, and is a useful check on the accuracy of calculations and drawing.

(ii) The angle between the two regression lines depends on the correlation coefficient, ranging from 0 (when $r = 1$ or -1) to $90°$ (when $r = 0$).

Consolidation
1. Construct a regression line for the data in Table 8.4 by the method of semi-averages. Plot the two least-squares regression lines on the same graph (using the equations given above) for comparison.

2. Convert the data in Table 8.4 into ranks, and plot the regression line and vertical residual lines in the way described in section 8.5a.

3. Calculate the least-squares regression equations for the hypothetical data in Table 8.3, and draw the regression lines.

8.6. CONFIDENCE LIMITS OF LEAST-SQUARES REGRESSION LINES
Unless two variables correlate perfectly, it is not possible to use regression lines to predict with complete accuracy and reliability. It is possible, however, to estimate with a given level of confidence, the range within which a value will lie, in the same way that confidence limits are calculated for estimates of population parameters (described in Chapter Five). This is done by using what is called the standard error of estimates made from the regression line, and is given by the formulae:

$$\text{S.E.}_{\cdot y \text{ on } x} = \sigma_y \sqrt{(1 - r^2)} \text{ for estimates of } y \qquad \textit{Formula 8.6a}$$

$$\text{S.E.}_{\cdot x \text{ on } y} = \sigma_x \sqrt{(1 - r^2)} \text{ for estimates of } x \qquad \textit{Formula 8.6b}$$

The standard error of estimates is in practice equivalent to the standard deviation of residuals (except that the former relates to an infinite sampling distribution and the latter to a finite distribution of actual observations). Both are associated with distributions which, it is generally assumed, tend towards normal as sample size increases, so that about 68 per cent and 95 per cent of observations will lie within 1 and 2 standard errors of the regression line respectively. Therefore 68 per cent and 95 per cent confidence limits will be 1 and 2 standard errors respectively either side of the regression line.
    This is illustrated for estimates of y in Figure 8.8, where the standard error of estimates has been calculated by:

(i) Finding $\hat{\sigma}_y$, the best estimate of σ_y, by applying Bessel's correction to S_y : $\hat{\sigma}_y = 11\cdot5 \times \sqrt{(16/15)} = 12\cdot0$
(ii) Calculating $\hat{\sigma}_y \sqrt{(1 - r^2)}$: $12\cdot0\sqrt{(1 - (0\cdot64^2))} = 9\cdot2$.

The dotted lines are then drawn on the graph parallel to the regression line, at intervals of $9\cdot2$ y-units, as shown. It can now be said that the

95 per cent confidence limits for the estimate of y corresponding to, say, $x = 40$, are $63 \cdot 6 + 2 \times 9 \cdot 2 = 81 \cdot 8$ and $63 \cdot 6 - 2 \times 9 \cdot 2 = 45 \cdot 4$ — i.e. on the evidence of Table 8.4, 19 out of 20 British towns with 40 per cent of their work force in Industrial Orders I to XVI contained populations in 1951 in which between $45 \cdot 4$ per cent and $81 \cdot 1$ per cent left school before the age of 15. These very wide margins show that, in this instance, the regression equation is of little use for making estimates; it serves rather as a succinct statement about the average relationship between employment structure and educational opportunity in a sample of British towns in 1951.

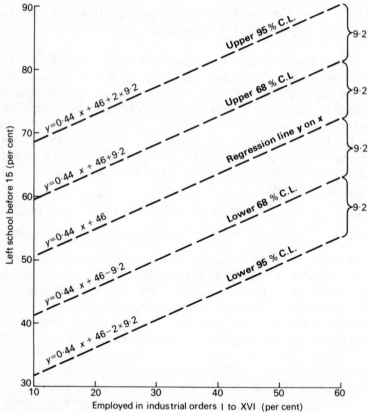

Figure 8.8. Regression line for y on x with 68 per cent and 95 per cent confidence limits for data in Table 8.4

Consolidation

4. (a) For the data in Table 8.4, calculate the standard error of estimates for the regression line of x on y.

(b) Plot the regression line (the equation of which has already been found to be $x = 0.92y - 15.5$ . . . see p.259), together with the lines of 95 per cent confidence limits.

(c) Identify any town located outside these confidence limits, and state precisely what this location signifies.

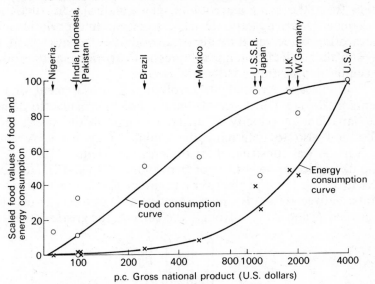

Figure 8.9. The relationship between p.c. GNP and food consumption (o) and energy consumption (x) for eleven major countries (late 1960s)

8.7. REGRESSION FOR NON-LINEAR RELATIONSHIPS

Figure 8.9 shows the relationship between p.c. GNP (on a log scale) and scaled values of (i) food consumption, and (ii) energy consumption, taken from Table 1.12 (p.41). Regression curves have been sketched in by eye, suggesting that as p.c. GNP increases:

(i) food consumption at first tends to increase rapidly and then more slowly as 'saturation level' is reached;

(ii) energy consumption at first tends to increase slowly and then more rapidly.

The difference in elasticity of demand implied by the different shapes of these curves (which could be replicated by many other social and economic indicators) would be obscured by linear regression lines: curves roughly sketched by eye give a much better idea of how the variables are related, and yield better predictions than the most precisely constructed least-squares straight regression lines. They would yield even better predictions (in the statistical sense) if the curves were constructed mathematically, though whether they would be better predictions in the sense of forecasting the future is doubtful, as cross-sections of the present (which is what these curves represent) are notoriously unreliable guides to the future. It is therefore probably not worth the trouble to do more than sketch the regression curves in this case.

On the other hand, where maximum accuracy of prediction of one variable from the other is required, mathematical construction of regression lines is necessary. In the case of non-linear relationships, this can often be achieved by first transforming the data logarithmically, and then calculating linear regression equations, as the following example demonstrates.

A field exercise for students involved the measurement of river currents at different distances from the bed and banks, to demonstrate variations in velocity at different points in the river's cross-profile, and also to determine the volume of discharge. A simple current meter was constructed for the purpose (Figure 8.10).

The difference in height (h) of two columns of paraffin (A and B) results from the force of the river current at C compressing the air in the tube between C and B; and the suction at D causing expansion of air between A and D: the stronger the current, the greater the differ-

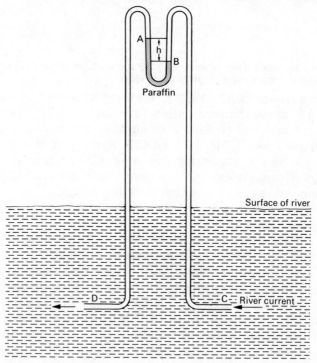

Figure 8.10. A home-made current meter

ence in height. The problem in which regression was used was one of calibration—to relate the difference in height of the paraffin columns (h) (measured in tenths of an inch) to the velocity of the current (v) (measured in metres per second). A sophisticated (and expensive!) current meter was borrowed; readings were taken at different points in a non-turbulent reach of the River Trent, using both home-made and borrowed instruments at each point; and the readings were plotted on to a scattergram. Although only seven pairs of readings were taken (the task was not as easy as it may seem), they yielded a product moment correlation coefficient of +0·98, which is significant at the 0·001 level. But the scattergram revealed a relationship which appeared parabolic rather than linear—see Figure 8.11.

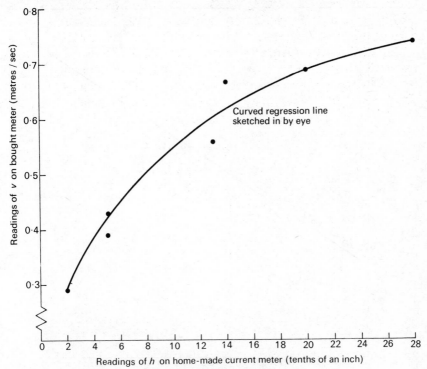

Figure 8.11. Scattergram to show the relationship between readings on two current meters

Both sets of data were therefore transformed by substituting logarithms for the numbers themselves, as set out in Table 8.5. The 1̄ prefixes to the values of log V, obtained from log tables, were eliminated by substituting log $10V$ for log V. The values of log h were called x and those of log $10V$ were called y. When these were plotted on a scattergram, a strong linear relationship was revealed (Figure 8.11).

TABLE 8.5
Logarithmic transformation of current meter readings

h in tenths of an inch	V in metres per sec.	log h = x	log V	log 10V = y
2	0·29	0·30	$\bar{1}$·46	0·46
5	0·39	0·70	$\bar{1}$·59	0·59
5	0·43	0·70	$\bar{1}$·63	0·63
13	0·56	1·11	$\bar{1}$·75	0·75
14	0·67	1·15	$\bar{1}$·83	0·83
20	0·69	1·30	$\bar{1}$·84	0·84
28	0·74	1·45	$\bar{1}$·87	0·87

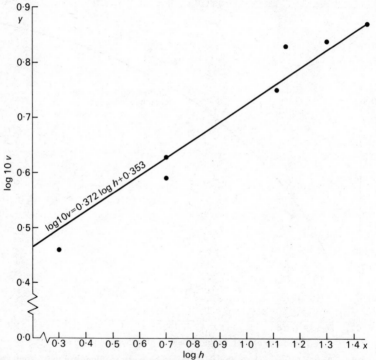

Figure 8.12. Scattergram and regression line on logarithmic scales, to show relationship between readings on two current meters in Table 8.5 .

Therefore a regression equation for y (i.e. log 10V) on x (i.e. log h) was calculated (since it was required to find values of V from values of h).

The resulting equation was:

$$y = 0\cdot372x + 0\cdot353$$

i.e. $\log 10V = 0\cdot372 \log h + 0\cdot353$

The line representing this equation has been drawn on the scatter-gram (Figure 8.12), and is seen to be a very good fit. One could now read values of y for different values of x from the graph. But more accurate figures are obtained by direct substitution in the equation. The first few rows of a table of calculations are shown below; when this was complete, the first and last columns of the table were stuck on to the home-made current meter, so that the differences in height of the paraffin columns could immediately be translated into velocity readings.

h tenths of inch	$\log h\ (=x)$	$0\cdot372x$	$0\cdot372x + 0\cdot353$ $(= \log 10V = y)$	Antilog y $(= 10V)$	V (metres/ sec)
1	0·0000	0·	0·353	2·25	0·225
2	0·3010	0·112	0·465	2·92	0·292
3	0·4771	0·177	0·530	3·39	0·339
etc.	etc.	etc.	etc.	etc.	etc.
↓	↓	↓	↓	↓	↓
50	1·6990	0·632	0·985	9·66	0·966

This particular example raises a more general question: how can we determine (other than by eye) whether a transformation of data will yield a better regression line from which to predict? And what kind of transformation will be best? Just taking logarithmic transformations, there are four possibilities.

(i) $y - mx + c$ (which can otherwise be written $y = f(x)$, meaning that y is a function of x—i.e. no transformation)

(ii) $y = m \log x + c$, or $y = f(\log x)$—only x is transformed.

(iii) $\log y = f(x)$—only y is transformed.

(iv) $\log y = f(\log x)$—both x and y are transformed.

It might be thought that a simple way of resolving this question is by calculating r (the coefficient of linear correlation) in each case, since r is a measure of the goodness of fit of data on to a least-squares straight regression line. But r measures the goodness of fit in terms of the variables in the equation—e.g. the regression line $\log y = f(x)$ minimizes residuals of $\log y$; and r measures the goodness of fit in terms of $\log y$, which is not necessarily the same as the goodness of fit in terms of y. Therefore, if values of y are to be predicted, it is the regression line which minimizes residuals of y, not of $\log y$, which is required.

The only way to determine the best transformation is to sum the squares of the residuals of the variable to be predicted, from each transformed equation in turn—a tedious operation without a computer, but straightforward and quick with a computer that has been suitably programmed. This was done with the current meter data, and a double log transformation ($\log V = f(\log h)$) gave the best fit— i.e. minimized the sum of the square of residuals of V (not $\log V$) on to the regression line. Hence this equation was used to determine values of V from corresponding values of h.

9
TESTS FOR DISTRIBUTIONS
IN SPACE

9.1. THE CONCEPT OF SPATIAL RANDOMNESS

ESSENTIALLY geography is concerned with distributions in space and one of the most important distributions the geographer has to consider is that of human settlement. For example, for many years the pattern of population tended to be described in terms of 'dense' or 'sparse', whereas settlement was either 'nucleated' or 'dispersed'. Recently considerable attention has been given to devising a more precise mathematical way with which areal distributions can be described. (Certain aspects have already been discussed in 2.3biii above.) Much of the pioneer work in geography has been done by M. F. Dacey.

Basically, three extreme types of distribution can be distinguished: 1. uniform (or regular); 2. random, 3. clustered (or aggregated). These are shown in Figure 9.1. The points may represent settlements, or the positions in two-dimensional space of any other set of phenomena which can be regarded for practical purposes as being located at points. If the pattern is uniform the distance between any one point and its nearest neighbour will be approximately the same. If the pattern is clustered then there will be one or more groups with a

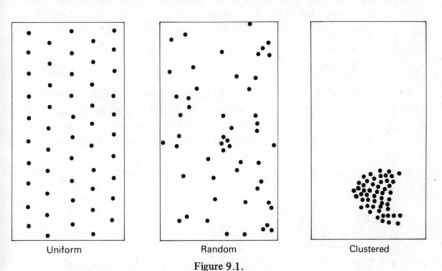

Uniform   Random   Clustered

Figure 9.1.

270 QUANTITATIVE TECHNIQUES IN GEOGRAPHY

relatively short distance between each point and its nearest neighbour, and large areas in which no point is located. If the pattern is random a situation results in which there is some tendency to cluster but also a number of points fairly widely distributed.

The problem was to find a single index for any given pattern, running on a continuous scale, from one extreme, when all points are clustered, to the other extreme, a situation in which all points are distributed uniformly over the whole area. One solution, originally devised by botanists seeking to describe plant distributions, is now widely used by geographers and is called the 'nearest neighbour index'.

9.2. THE NEAREST NEIGHBOUR INDEX

The importance of the index is that it provides a test for 'non-randomness' and allows, on a continuous scale, comparisons to be made of two or more spatial distributions. Settlement patterns have been chosen as an actual example, but the technique can be applied to many distributions with which the geographer is concerned, e.g. a particular manufacture, or retail or other service in an urban area. The index ranges from 0 (indicating that all points are closely clustered) to 2·15 (indicating that all points are uniformly distributed throughout the area, and therefore as far away from each other as possible). A random distribution is indicated when the index value approximates to 1.

The index value, normally written as R, is calculated simply enough by dividing the measured mean-distance between nearest neighbour points observed in a given area ($\bar{D}_{obs}$), by the mean distance to be expected from a similar number of points randomly distributed in the same area ($\bar{D}_{ran}$).

Or $R = \bar{D}_{obs}/\bar{D}_{ran}$ *Formula 9.1*

To calculate $\bar{D}_{obs}$ it is necessary to measure the distance between each point and its nearest neighbour and divide by the total number of measured pairs. (A point may sometimes be included more than once as it can form the nearest neighbour of more than one other point.)

$\bar{D}_{ran}$ is obtained from Formula 2.2, previously used in section 2.3biii,

$$\bar{D}_{ran} = \frac{1}{2\sqrt{\left(\frac{N}{A}\right)}}$$

where N is the total number of points and A is the given area.

Therefore

$$R = \frac{\bar{D}_{obs}}{1 \div \left(2\sqrt{\frac{N}{A}}\right)}$$

which may be simplified to $R = 2\bar{D}_{obs} \sqrt{/\left(\frac{N}{A}\right)}$ *Formula 9.2*

Note. The unit of measurement is immaterial provided the same unit is used throughout. For example, if distances are calculated in kilometres, then the area must be in square kilometres.

An example will show clearly the meaning of R, and how the index is obtained.

Figure 9.2a shows the location of villages in a rural area south of Scarborough. From inspection, village settlement appears fairly well scattered, although there is some indication of a linear pattern in part of the area. For purposes of comparison it is decided to calculate the nearest neighbour index for the area.

$$R = 2\bar{D}_{obs} \sqrt{/\left(\frac{N}{A}\right)} \quad \text{or} \quad R = \frac{\bar{D}_{obs}}{1 \div \left(2\sqrt{\frac{N}{A}}\right)}$$

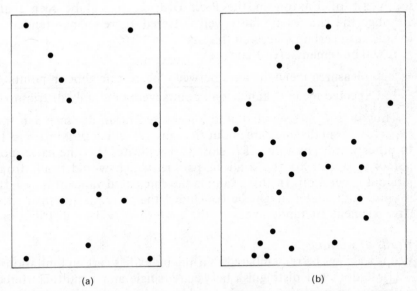

(a)   (b)

Reduced from the one inch O.S. maps from which the actual calculations were made

Figure 9.2. (a) The location of villages in a rural area south of Scarborough
(b) Villages in an area of the Peak District south-east of Buxton

where $\bar{D}_{obs}$ is the mean distance between the nearest pairs of observed points, N is the total number of observed points, and A is the observed area.

In this example

$$R = 2 \times 1\cdot45 \times \sqrt{\left(\frac{16}{60}\right)} \quad \text{or} \quad R = \frac{1\cdot45}{1 \div 2 \sqrt{\left(\frac{16}{60}\right)}}$$

$$= \underline{1\cdot50}\ldots \qquad\qquad = \frac{1\cdot45}{0\cdot968}$$

$$= \underline{1\cdot50}\ldots$$

where $\bar{D}_{obs} = 1\cdot45, N = 16, A = 60$, and $\bar{D}_{ran} = 0\cdot968$.

We now have a precise statistic, $R = 1\cdot50$, to describe the degree of 'scatter' of the villages in the area under review. Remembering that the 'R' scale runs from 0 (values approaching 0 indicate close clustering) to 2·15 (an even distribution over the whole area) we have an index to show how close to uniformity the distribution of settlement is, and statistical evidence that it is most unlikely this pattern developed purely through 'chance'.

Figure 9.2b shows the distribution of rural settlement in an area south-east of Buxton in the Peak District. It will be seen that, although the villages are fairly well scattered, there seems a tendency to cluster in certain places. In this case $R = 1\cdot10$.

It will be remembered that

$$R = \frac{\text{Measured mean distance between 'nearest neighbour' points.}}{\text{Expected mean distance if all points were randomly distributed.}}$$

Thus it can be seen that if the measured mean distance and the expected mean distance are about the same, R will approximate to 1. In other words any value of R close to 1 indicates that the pattern of points is similar to the kind of pattern which would result from random generation. In this example the calculated value of $R = 1\cdot10$ is close to 1 and it might be concluded that there is an apparently large element of randomness in the pattern of village distribution.

Words of warning
It must always be remembered that the index has certain limitations:
(i) It does not distinguish between a single and a multi-clustered pattern. (An extreme example of this might be a series of pairs of towns on either side of a river, each pair at some distance from the next, making a linear pattern in the landscape. Because each town

forms the nearest neighbour of the other town of the pair the index will be near 0.)

(ii) It 'averages out' sub-patterns which may exist within the area, and may thereby hide contrasting sub-patterns which cancel each other out when put together to give a false impression of randomness.

(iii) The selection of the nearest neighbours is arbitrary (and time-saving). The choice of the second or third nearest neighbour is possible and might produce different results.

(iv) The fact that a pattern emerges with an index approaching 1 indicates the kind of distribution which could result from the independent random selection of each location. This does not mean that the locations of the observed phenomena in the real landscape are *necessarily* purely the result of 'chance'.

(v) The distribution of one kind of phenomenon may largely determine that of another. For example, a particular type of drift deposited by melting ice in an apparently random manner might closely affect a pattern of agricultural land use.

The two great advantages of the nearest neighbour index are that it enables more exact comparisons to be made than hitherto, and it demonstrates clearly point patterns that are *not* random, and which therefore require explanation. Like all statistical methods it has to be applied with caution and common sense. The value of the interpretation of the results depends on the judgement and skill of the investigator.

So far R has been used in two ways. First, as an index to *describe* the extent to which a two dimensional distribution approaches complete nucleation or completely even distribution, and secondly to determine how closely it resembles the randomly generated distribution of the same number of points in the same area. However, the index is not only descriptive but also allows an inferential test, and therefore it is necessary (as in any other inferential test) to calculate the significance of the result. In other words, having obtained an index for a particular distribution it is necessary to calculate the probability of such a distribution resulting from 'chance'. Unless the probability is acceptably low, whilst R may be used *descriptively*, no conclusion can safely be drawn based on the value of R alone that the distribution is not the result of random variations.

Fortunately a simple test is described by L. J. King (1969, p.100), which gives an exact probability through the calculation of a z-score. First it is necessary to calculate the standard error of $\bar{D}_{ran}$, which is given by

$$\sigma_{\bar{D}_{ran}} = \frac{0{\cdot}26136}{\sqrt{\{N \times (N/A)\}}} \qquad \textit{Formula 9.3}$$

where N is the number of points in the distribution and A is the area in which they are located.

The value of z may then be obtained by

$$z = \frac{\bar{D}_{obs} - \bar{D}_{ran}}{\sigma_{\bar{D}_{ran}}}$$ *Formula 9.4*

Let us now test for significance the statistic $R = 1\cdot50$, that we obtained as a descriptive index for the village pattern shown in Figure 9.2a.

1. H_0 is that the pattern of villages is similar to a pattern produced by the independent random location of each point.

2. H_1 is that the villages are distributed in a manner that is not random.

3. Level of rejection is decided at $\alpha = 0\cdot01$.

4. We first calculate the standard error of $\bar{D}_{ran}$ from Formula 9.3.

$$\sigma_{\bar{D}_{ran}} = \frac{0\cdot26136}{\sqrt{\{N \times (N/A)\}}}$$

$$= \frac{0\cdot26136}{\sqrt{\{16 \times (16/60)\}}}$$

$$= 0\cdot126 \ldots$$

It is now possible to obtain a z-score by the use of Formula 9.4.

$$z = \frac{\bar{D}_{obs} - \bar{D}_{ran}}{\sigma_{\bar{D}_{ran}}}$$

$$= \frac{1\cdot45 - 0\cdot968}{0\cdot126}$$

$$= \underline{3\cdot81} \ldots$$

5. With a value for $z \geqslant 3\cdot5$, Appendix 1, col. C shows that the associated probability $\simeq 0\cdot000$ (for a two-tailed test). That is, this distribution might be expected to occur through chance variations less than once in a thousand, a very high level of significance. The geographer is thus justified in rejecting H_0 and in seeking explanations to account for such an even settlement pattern.

On the other hand the pattern of villages in part of the Peak District (Figure 9.2b) has a calculated $R = 1\cdot10$, showing that it approximates to a randomly generated distribution. A z-value of $0\cdot937$ tends to confirm this; which means that the pattern revealed is close to the kind of distribution which would occur if each point were independently located, and the result of 'chance'. But it must be remembered that this is only a *mathematical* relationship. There is good reason to believe in the case of this example that the slight clusters and the blank areas in the pattern, although to be expected in a random distribution, are partly the result of the influence of physical factors. Nearest neighbour analysis may be used to prove that a distribution is significantly *not* random; it cannot be used to prove that it is. If the analysis shows randomness in the statistical sense, confirmation must lie outside measurements and figures, and within the judgement of the investigator.

The boundary problem

In any calculation of R the value of N may be determined with precision and, provided care is taken, so may the value of D_{obs}. This does not apply in the same way to A, other than in exceptional cases. For example, if the survey includes a whole island, the use of high-water mark puts an exact limit to area. Similarly, the distribution of a plant species over a particular rock type has a natural boundary when the outcrop terminates. But we are rarely fortunate enough to be in this kind of situation and are more often faced with the problem of exactly where to draw our boundary line (ignored for the sake of simplicity in Figure 9.2).

For the purpose of nearest neighbour analysis distributions may be considered as of two kinds. They may, like the pattern of settlement, extend across the whole landscape, requiring an arbitrary decision to be made concerning the boundary of the sample area, or they may be regarded as having 'natural' limits, for example, a particular urban attribute, such as the study of a certain type of retail service, distributed throughout one town. In this case the urban fence, or the edge of the built-up area, may provide an acceptable boundary line because the observed pattern is defined specifically as a function of the urban area. But the solution is not always so easy. If we seek to test a distribution within the core (or C.B.D.) of a town the difficulties in defining an exact area are considerable. (A very good analysis of the problem is given by H. Carter, *The Study of Urban Geography*, 1972, pp.193–211.)

There is no ready-made solution to the boundary problem. Fortunately, where an acceptable boundary does not already exist there are sometimes methods which can be used to define one for the purpose of the test. The method of construction will vary according to the problem and will generally have to be devised to suit each different situation and the type of phenomena involved. For example, suppose we wish to calculate R for the distribution of settlement in Norfolk and Suffolk (Figure 9.3). The sea coast provides an effective natural boundary to the east and north; but the southern and western boundaries are formed by the administrative borders of the counties. The question is: 'Do the county borders provide an appropriate limit for the calculation of area?' And the answer must be that they do not: a decision which is not based on any mathematical concept but on general geographical knowledge, and arising out of the nature of the phenomena with which we are dealing.

The difficulty is this. The statistic, R, which we wish to obtain, will be a measure of the evenness or concentration of the settlement pattern in both counties. In this connection it must be remembered that part of the pattern is the rural countryside itself which separates towns and villages. In it some settlements may to a varying degree be interdependent. There also exists a relationship between each settlement and the countryside around it. But county boundaries, hardly altered since they were first created in the Middle Ages, rarely reflect the present influence of individual settlements close to the border, which often extends into an adjacent county. Thus some of the space (or area) forming part of the pattern lies outside the county boundary. For example, much of the southern border of Suffolk is formed by the River Stour, along the north bank of which a line of settlements has developed. If the area under consideration were to be limited in the south strictly to the county boundary, some of the area—functionally part of the settlement pattern of Suffolk—would be excluded from the test.

In this particular case there appear to be at least two possible solutions. The first possibility is to mark the mid-point between every point nearest the border within Norfolk and Suffolk and that of the nearest point in the neighbouring counties, and join these by eye in the form of smoothed curves to make a new boundary to include small parts of adjacent counties and lose a little of the study area. The second might be to follow the same procedure but take the mid-point between the last two points on the Norfolk/Suffolk side of the border, sacrificing a few points and a little area, but confining the test strictly within the county boundaries. In both methods towns outside the boundary are excluded from the calculation. Either seems a reasonable and objective way to construct a realistic boundary for the estimation of A. Both methods are shown in Figure 9.3. It should be noted that the larger the area chosen and the greater the number of points, the smaller will be the result of decisions affecting the boundary. In the above example for the smaller area $R = 1 \cdot 37$, and for the larger area $R = 1 \cdot 38$—a negligible difference as both are statistically highly significant.

In cases where a very large number of points are involved it may

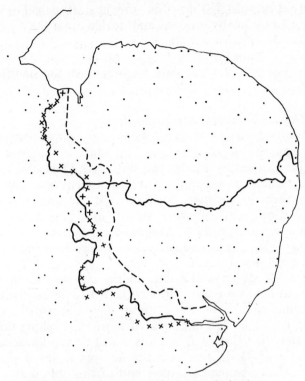

+ + + Median line between every point nearest the border
and that of the nearest point in neighbouring countries.

– – – Median line drawn between the last two points
on the Norfolk/Suffolk side of the border

Figure 9.3. Settlements in Norfolk and Suffolk as shown in *The University Atlas*, 7th edn.,
edited by Goodall and Darby. Every settlement, irrespective of size, is represented by a point

be desirable to estimate $\bar{D}_{obs}$ by sampling. But care must be taken to
ensure that the sample is really random. The value of the sample
$\bar{D}_{obs}$ may then be used in the calculation, together with the total
number of points (N) within the boundary, whether a part of the
sample or not.

Summary
1. Calculate the area in which the points are located (A), and count
   the number of points (N).
2. Measure the distance between each point and its nearest neighbour. (Remember that two points may be their own nearest
   neighbours, in which case the distance between them is counted
   once for each point.)
3. Calculate the mean of these observed distances ($\bar{D}_{obs}$).

4. Apply Formula 9.2 to find the nearest neighbour index (R).
5. Formulate your hypotheses and decide upon the rejection level to test the descriptive statistic R.
6. Find $\sigma_{\bar{D}_{ran}}$ from Formula 9.3.
7. Using Formula 9.4 calculate z to establish the significance of the result.

9.2a. The linear nearest neighbour index

An interesting variant of two-dimensional nearest neighbour analysis is the application of the same principle in a linear situation, a technique devised by Pinder and Witherick (Tijdschrift, 1973). This method, which can be applied to a unidimensional distribution, was used by them (*Geography*, 1975) to test the degree of clustering of clothing shops, and of chain stores, in a busy retailing area. The technique can obviously be adapted for other purposes, for example, with any series of discrete categories spaced in a linear sequence, as in a conventional transect.

Figure 9.4 displays the data in the chosen example, each shop being regarded as a point. The first task is to define the length (L) of the retailing ribbon. In doing this we meet the same kind of problem that we found in defining area, that of establishing limits, because the greater the value of L the lower will be the linear nearest neighbour index (LR). In this case the authors believed that a 'natural' limit could be assumed at either end of the ribbon at a point where the continuity of shops gave way to dwellings. Obviously criteria for determining the length of L will vary with circumstances. In some places it may be simple and obvious, in a New Town for instance. Generally in our older cities it will be a more difficult matter of judgement and common sense. One method, used by the Census of Distribution to delimit a central shopping area, is to place the boundary at the end of the block containing less than one-third shops.

$\bar{D}_{obs}$ may be calculated in two ways. One way is to measure the actual linear distance along the ribbon between the centre of every shop front and its nearest neighbour, and to find the mean distance for all the points in the same way that $\bar{D}_{obs}$ was determined for a two-dimensional distribution. The second way assumes that the relationship between different shop units is more important than actual distance. The length of shop frontage is therefore disregarded and $\bar{D}_{obs}$ is calculated in terms of the number of shops intervening between each pair of nearest neighbours. Both methods are mathematically acceptable. The second was used here as it seemed more appropriate in this kind of investigation.

There remains the task of finding a linear form of $\bar{D}_{ran}$, that is, of

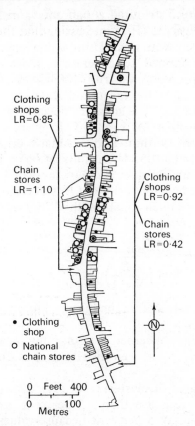

Clothing
shops
LR=0·85

Chain
stores
LR=1·10

Clothing
shops
LR=0·92

Chain
stores
LR=0·42

• Clothing
shop
○ National
chain stores

0   Feet   400

0   100
     Metres

Figure 9.4. The location of clothing shops and national chain stores along the retailing ribbon of North End, Portsmouth

determining the mean number of units (or distance) between the same number of points randomly distributed along a line of equal length. It will be remembered that the two-dimensional mean is found from the formula $\bar{D}_{ran} = \dfrac{1}{2\sqrt{N/A}}$, which may also be written $\bar{D}_{ran} = 0.5 \sqrt{\dfrac{A}{N}}$. The authors have found experimentally that by using a modified formula

$$L\bar{D}_{ran} = 0.5 \left(\dfrac{L}{N-1}\right)$$

Formula 9.5

where L is the line length, and N the number of points in an observed linear distribution, a very close approximation is obtained of a similar number of randomly distributed points. It will be seen that it so happens that, by chance, the mean distance between nearest neighbour points randomly distributed is half the value it would be if the points were evenly spaced. (It is interesting that the confirmation of this formula is entirely empirical. Using a line length of 1000 units a computer was programmed to generate a series of random point distributions. The program was run 100 times at intervals of ten between 10 and 100, and the mean and standard deviation of every set of 100 runs calculated. The standard deviations were then used to obtain the standard error for each set. In every case the mean lay within–generally well within–two standard errors of the mean predicted from Formula 9.5. The results of using Formula 9.5 may thus be accepted as significant at the 0·05 level.)

The linear nearest neighbour index LR is obtained in a similar way to that of the two-dimensional method, by obtaining the ratio between $L\bar{D}_{\text{obs}}$ and $L\bar{D}_{\text{ran}}$:

$$LR = \frac{L\bar{D}_{\text{obs}}}{L\bar{D}_{\text{ran}}} = \frac{L\bar{D}_{\text{obs}}}{0 \cdot 5 \left(\dfrac{L}{N-1} \right)} \qquad \textit{Formula 9.6}$$

It is now necessary to define the inquiry more precisely by formulating our hypotheses:

H_0 is that the shops concerned are randomly located, and that any indication of clustering, or even distribution, is due to chance variations.

H_1 is that the degree of clustering, or even distribution, is so great that we can say, there is less than a 5 per cent probability it is the result of chance. (We say 5 per cent because values for LR are given in Figure 9.5 only for the 0·05 level of significance. Hence α must equal 0·05.)

It is also important to remember that, although we speak of

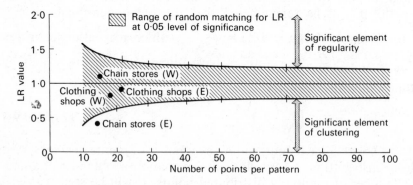

Figure 9.5. The 0·05 level of significance of LR for values of $N = 10$ to $N = 100$

random locations, it is most unlikely that the position of any shop will ever be solely the result of chance. Factors will have been taken into account in its location, some of which may be unascertainable. These are numerous and may operate to produce a result similar to that of a random distribution. This may comprise clusters and even distributions which, when averaged out over the whole transect, create a random effect. As with the two-dimensional analysis, if the result is significantly *non*-random we know it is unlikely to be due to 'chance'; but if it falls within the random area of Figure 9.5 we cannot make the opposite inference.

As previously, values of LR close to $1\cdot0$ indicate that the observed distribution is similar to the same number of points randomly distributed along a line of similar length. A test at the $0\cdot05$ level has been worked out by the authors for different values of N to show when LR is significantly different from $1\cdot0$, and is reproduced in Figure 9.5. The area above the shaded portion of the figure indicates significant regularity, and that below significant clustering. If the points are all adjacent $LR = 0$, if perfectly evenly distributed $LR = 2\cdot0$. With $N < 10$ the test is unreliable.

It will be seen that values of LR for clothing shops and chain stores have been indicated in Figure 9.4 separately for either side of the retailing ribbon. This is because the road is a very busy thoroughfare, and each side may be considered independent to some degree. When access across a road is easy, as in a pedestrian precinct, both sides would have to be considered as one line.

The values of LR have also been plotted on Figure 9.5, and it will be seen that the only statistically significant result is the clustering of chain stores. Explanation of the results of linear nearest neighbour analysis lies in the field of urban geography. But if significant values of LR emerge, they may provide useful indications of the directions in which further research may be profitable, and a challenge to the urban geographer in the exercise of his craft.

The work of Pinder and Witherick is an interesting example of the progress that can be made in quantitative methods by geographers using a computer, with no more mathematical expertise than is contained in this book.

Summary
1. Define the line length (L) in which the points (N) are located.
2. Measure the distance between each point and its nearest neighbour, using either method described above, and calculate the mean $(\bar{D}_{obs})$.
3. Formulate hypotheses.

4. Use Formula 9.6 to calculate *LR*.

5. Ascertain from Figure 9.5 whether *LR* is significant.

Consolidation

1. Calculate the nearest neighbour index for the chip shops within the 1 mile circle given in Figure 2.5, and test *R* for significance. Compare your results with those of other techniques outlined in Chapter Two.

9.3. A SIMPLE RANDOMIZATION TEST

A number of methods have been devised to test whether or not a *sequence* may be considered random. For example, suppose a queue of people had formed for admission to a popular event that was equally attractive to men and women. If the order in which the individuals arrived to take their place in the queue was uninfluenced in any way, that is if the arrival of a man or a woman was determined by chance, then the position of men and women in the queue would be random. We should expect to find small numbers of men and small numbers of women standing behind each other in some places. In others men and women would be placed alternately. The order in the queue might look something like this:

(a) **MMWMWWWMMWMWWMMMMWWWMWWM**

But if the queue consisted entirely of local workers, and suppose the women finished work earlier than the men, because the women had a time advantage over the men the order might take an even distribution which is clearly far from random, thus:

(b) **MMMMMMMMMMMMWWWWWWWWWWWW**

If, on the other hand, for some reason the event was restricted to married couples, irrespective of when they arrived the order of men and women (ladies first!) would be:

(c) **MWMWMWMWMWMWMWMWMWMWMWMW**

A different kind of even distribution, but again one with a very small probability of occurring by chance.

Now there are 12 men and 12 women in each of (a), (b), and (c) above, and when we speak of a 'random' distribution we are referring to the *sequences* in which they appear. A single person or any number of men or women standing together constitute a 'sequence', more usually called a 'run'. Let us replace M by +, and W by −, and rewrite (a), (b), and (c), recording cumulatively the number of runs in each.

(a) + + − + − − − + + − + − − + + + + − − − + − − +

runs 1   2 3   4   5 6 7 8   9   10 11 12 13

(b) + + + + + + + + + + + + − − − − − − − − − − − −

runs   1   2

(c) + − + − + − + − + − + − + − + − + − + − + − + −

runs 1 2 3 4 5 6 7 8 9 10 11 12 13 14 15 16 17 18 19 20 21 22 23 24

We have agreed that (a), which has 13 runs, seems to represent the 'chance' arrival of men and women to join the queue. But the 'chance' arrival in (b) of all the women first, and in (c) of a man and woman alternately, seem both equally improbable, and in our fictitious example reasons have been suggested to account for this.

It may thus be seen that the number of runs in a series may be too few, or too many, to be a likely result of 'chance'. It is possible, however, to calculate the mathematical probability for any number of runs in a given series occurring by 'chance'. Appendix 11A is arranged to show the smallest number of runs likely to occur with a probability of 5 per cent or less (i.e. significant at the 0·05 level) through chance variations for samples having pluses (n_1) 2 to 20, and minuses (n_2) 2 to 20. Appendix 11B gives the same probability for large numbers of runs. In the examples (a), (b), and (c) above with 12 pluses (n_1) and 12 minuses (n_2), runs of 7 or less, or 19 or more have a 5 per cent or less probability of occurrence through 'chance'. (b) with 2 runs, and (c) with 24 runs are thus statistically significant, and (a) with 13 runs is apparently random.

This method of establishing probability may be applied to spatial distributions, providing it is possible to reduce these to an 'either/or' linear sequence without altering the natural order in which they occur. Figure 9.6 represents a fictitious distribution of the kind that might be found, for example, on a land-use map. (In the case of a map the lines of the National Grid could be taken to represent the rows.)

1. Our null hypothesis, H_0, is that the variations from north to south in the hypothetical distribution are due to chance.

2. The alternative hypothesis, H_1, is that the variations from north to south are significantly different from variations which might be expected from the independent random location of each point.

3. The rejection level α is 0·05.

4. The number of occurrences in each row is totalled and recorded, and the median found. In this example, the value of the median is 4. We allot a + to each row above the median value, and − to each row below. It is then possible to set these out in a linear

sequence, and count the number of runs:

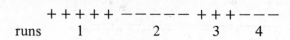

5. Reference to Appendix 11 shows that for a series in which $n_1 = 8$, $n_2 = 8$, the critical number of runs that have a 5 per cent probability or less of occurring through chance variations is 4 or less, or 14 or more. In our example there are 4 runs and we can therefore say that this is statistically significant at the 0·05 level, and reject H_0. We may conclude that our spatial distribution varies significantly in the *same direction* that we have made our *linear sequence*; in this case from north to south. For example, if the dots in Figure 9.6 represented all the farmhouses (or arable fields) in a certain area we could conclude that their distribution from north to south differed significantly from what could be expected from chance variations. It does *not* necessarily apply to variations from west to east *in the same distribution*.

The technique may be used to test any distribution that can be represented in a two-dimensional plane, and adapted to suit differing situations. For instance, the dots considered above were regarded as point locations, but if they represented areas which differed considerably in size a system of weighting could be introduced in which the value of the dot was proportional to its actual area on the ground (a modification which might make the result of considerably more value if an aspect of rural land use were being considered). In this example the median was used to dichotomize the data. In this or different situations the mean could be used, or some other criterion which was more suitable in the judgement of the investigator. The method is useful because it is simple and can be completed so quickly, especially if the distribution to be tested is already on a map with a superimposed grid.

The use of the runs test with data recorded in the form of a transect has obvious advantages. It may be applied whenever the data can be converted to an 'either/or' nature. It might, for example, be employed to test for randomness the distribution of a particular species along a line of transect across a salt marsh; or the distribution of some phenomenon, such as size or type of shop in an urban area. (Although the result is expressed in linear form there is no reason why the transect itself should necessarily be straight, provided that the data can be regarded as a continuous series.)

One of the limitations of the method described above is that the number of pluses and minuses (n_1 and n_2) that can be used with

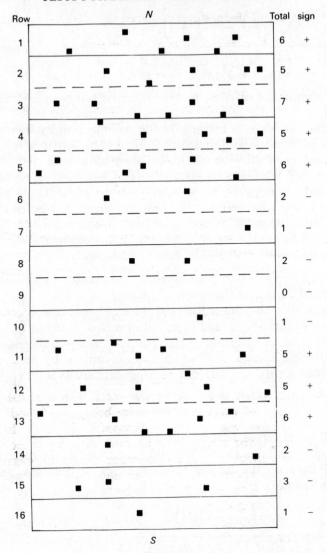

Figure 9.6

Appendix 11 is each limited to 20. However, if either n_1 or n_2 exceeds 20 it is permissible to calculate a z-value and determine probability from the normal distribution table. The z-value can be obtained from the relatively simple (though rather bulky and formidable-looking) formula:

$$z = \frac{r - \left(\dfrac{2n_1 n_2}{n_1 + n_2} + 1\right)}{\sqrt{\left(\dfrac{2n_1 n_2 (2n_1 n_2 - n_1 - n_2)}{(n_1 + n_2)^2 (n_1 + n_2 - 1)}\right)}} \qquad \textit{Formula 9.7}$$

where r = number of runs, n_1 = number of pluses, n_2 = number of minuses.

Suppose we wish to investigate the colonization by marram grass of a line of small sand-dunes on the edge of an estuary. Some of the dunes have grass growing on them, some have none. As a result of fieldwork we allot a plus to each dune with any grass (however little), and a minus to each dune with no grass. There seems no apparent cause why certain dunes are preferred, but before we seek reasons why the grass should take root on some dunes rather than others it is necessary to establish the statistical probability of the observed distribution being the result of 'chance'.

1. H_0 therefore is that the sequences represented by colonized and uncolonized dunes are the result of 'chance'.
2. H_1 is that the sequences are clustered, because vegetation on one dune will tend to spread to neighbours (a directional H_1).
3. The level of rejection of H_0 is decided at $\alpha = 0.05$ (one-tailed).
4. Allotting a plus for dunes already colonized by marram grass, and a minus for those which are not, our data are as follows:

$$+ + - - + - - - + - + + - + + + + - + + + + + + + + +$$
$$- - - + - - - + + + - - - - - - + + + + - + + + + +$$

It will be observed that of our 51 dunes 31 have a plus ($n_1 = 31$), and 20 have a minus ($n_2 = 20$), and that there are 19 runs ($r = 19$). We now have all the information necessary to obtain a z-score by applying Formula 9.7:

$$z = \frac{r - \left(\dfrac{2n_1 n_2}{n_1 + n_2} + 1\right)}{\sqrt{\left\{\dfrac{2n_1 n_2 (2n_1 n_2 - n_1 - n_2)}{(n_1 + n_2)^2 (n_1 + n_2 - 1)}\right\}}}$$

$$= \frac{19 - \left(\dfrac{2(31)(20)}{31 + 20} + 1\right)}{\sqrt{\left\{\dfrac{2(31)(20)[2(31)(20) - 31 - 20]}{(31 + 20)^2 (31 + 20 - 1)}\right\}}}$$

$$= \underline{-1.9} \ldots$$

5. With a value for z of $1 \cdot 9$ it will be seen from Appendix 1 col. B that the associated probability is $0 \cdot 029$. Therefore we may reject H_0 and accept H_1, with 97 per cent confidence. In the above example the z-value is a minus quantity. This does not affect the probability but is an indication that the number of runs involved is smaller than if they had been randomly distributed. A similar positive value of z would indicate the number of runs was too great to have more than a 3 per cent probability of being due to chance.

Summary
1. Decide whether it is possible to dichotomize data arranged in a natural series by assigning plus to one occurrence and minus to the other.
2. Formule H_0 and H_1, and decide on the rejection level.
3. Count the total number of pluses (n_1), and of minuses (n_2), and the number of runs (r).
4. If the sample is small establish whether r is great enough or small enough to be significant from Appendix 11.
5. If the sample is large with n_1 or $n_2 > 20$ apply Formula 9.7 and obtain the associated probability for z from Appendix 1 col. B or C.

Consolidation
2. Using the small sample method, and a land-use or topographical map, analyse a chosen pattern for directional significance.
3. As part of a study of immigrants in a Canadian town a transect was made outwards from the town centre by recording whether more than half the families in each block were immigrants. The result was as follows:

X X X X O X X O O X X X O X O X O O X O O O
X X O O X O O O O O O O O O O
    X   Immigrant
    O   Not immigrant

With what degree of confidence can it be stated that the blocks with $> \frac{1}{2}$ immigrant population are not randomly distributed along the line of transect?

MODELS AS QUANTITATIVE TECHNIQUES

10.1. INTRODUCTION

MUCH has been written about the use of models in geography, especially since about 1960, and no worthwhile brief summary of the whole field is possible. (There are now a number of standard texts available to which the interested reader may care to make reference). The object of this chapter is to form an introduction to some of the theory behind the making and use of quantitative models, and to indicate certain selected techniques which, while shedding light on the theory, may prove of practical value and use to the reader.

The word 'model' is used in two ways. The more embracing is that it is a purposeful 'simplification of reality'. In this sense, most of this book is concerned with models. The normal and exponential curves are models; so are regression lines, and so are statistics such as the mean, standard deviation and correlation coefficient. But a more restrictive definition of the word is 'the formal presentation of a theory' (D. Harvey, 1967), which implies some kind of explanation, prediction, or prescription. *Explanatory* or *predictive* models may be further differentiated as either *deterministic* (yielding precise and certain outcomes) or *stochastic* (yielding only approximate and probable outcomes). *Prescriptive* models are also called *normative*; they purport to determine what ought to be, rather than explain what is or foretell what will be.

Models are a simplification of reality. They are a method of extracting certain phenomena for study in which the (assumed) influence of other factors is eliminated, or compensated for in some way, so that we can achieve a better understanding within a limited field. In the case of the map the amount of detailed information that cannot be shown depends upon scale. What finally appears depends upon the judgement of the cartographer and the use to which the map will be put. If conclusions are to be drawn from the relationships that are shown, then the validity of these relationships will depend in part on the criteria used in the initial selection of data.

The success of a model as a predictive technique depends not only upon an accurate quantitative assessment of the present causes of the phenomena studied, and the weighting given to each, but also on judgement as to how these may change in the future, or completely new causal factors emerge. For example, suppose a model is to be

made to predict future traffic flows in a town to act as a basis for
planned road construction. Traffic density in towns, especially
during peak periods, has been increasing for many years, largely due
to the increasing number of private cars and vans; but peak traffic
density cannot be forecast solely from the national per capita growth
rate of vehicles. Many other factors are involved which have to be
quantified and built into the model. At a time of urban renewal
different kinds of development generate different amounts of traffic,
especially at peak hours. A higher proportion of office workers own
cars and use them at peak periods than any other major category of
urban worker. Therefore the building of large office blocks in city
centres will cause a predictable increase in traffic density in the
morning and evening. The construction of a theatre, which may
generate its own considerable traffic, will do so generally at off-peak
periods, whereas high rise blocks of flats for office workers built
adjacent to the city centre will tend to reduce the home/work traffic
rush hours. New developments in public transport, changing social
attitudes towards car ownership, all these and many other factors
must be expressed in quantitative terms, accurately assessed, and
correctly projected forward in time, before an accurate model can be
produced to predict traffic flows at different hours of the day, for
specific periods in the future.

The so-called gravity model (see 10.4) is a simple example of a
predictive model of this nature. The theory is that the interaction
between two centres of population, generally expressed in terms of
the movement of people, or goods, or services, is dependent basically
upon two factors. One, the number of people who live in each
place—a reasonable assumption. And two, the further apart the two
places are from each other the fewer will be those prepared to make
the journey, and the fewer goods and services will be exchanged,
because of the increased difficulty, cost, or length of time taken.
The model expresses this theory mathematically, so that it is possible
to predict the effect of population clusters and the 'friction of
distance' upon traffic flow. Conceptual models like this are
essentially the expression of a theory. The construction and use of
such a model is shown on p.303.

A model is thus a technique to simplify reality. No map can show
every detail of the landscape it represents. And no predictive model
can include *every* factor present in the real situation, even if these
could all be known. Theory determines which factors are regarded as
dominant, and these are incorporated into the model. In addition,
there are those variables which we ascribe to 'chance', and which are
often allowed for within the model by the use in some way of

random numbers. The success of any model depends upon the correct selection and evaluation of the facts it is constructed to represent. Failure to give correct weighting, or to recognize causal components, will result in the failure of the model as a representation of reality. If landscape elements omitted from the map are in fact necessary to an understanding of the relationships under investigation, then the conclusions drawn will be faulty. And if incorrect weighting is given in a model designed to predict some future state the answer will be wrong. Reality is immensely complex, and the dangers of over-simplification are great. A model can be a very useful technique, but it must be evaluated with caution, and care must be taken not to place too much reliance on a result simply because it is arrived at by a precise mathematical formula.

Very many different kinds of model are used in geography. It is possible here to demonstrate only very few of the model-making techniques which now exist. Those selected for inclusion in this chapter were chosen for two main reasons:

a. They are all quantitative (not all theoretical models are necessarily quantitative).

b. They are models which actually *generate* statistical data. This may be in order to test the theory underlying the model by comparing the data it generates with real data observed in the field (of which the random walk technique used in 10.2 is an example). Or, like the diffusion technique in 10.3, the model may be used as a basis to predict the course of some future event or events.

10.2. SIMULATION MODELS–A RANDOM WALK TECHNIQUE

All models simulate, or represent, some aspects of reality. In geographical literature the term 'simulation' is usually taken to refer to those models in which *process* is the most important component. A very simple kind of simulation model is the creation on paper of a drainage pattern by the 'random walk' technique. Squared paper is used. Each of the four points of the compass is allotted a number. Every square in turn is then referred in sequence to a table of random numbers, which determines the direction in which the 'stream' will flow from that square. If the random number indicates an impossibility (e.g. a stream flowing back up its own channel, or an area of circular drainage) that number is ignored and the next one chosen. The network generated in this way may be made entirely random, by giving each direction an equal weighting, or it may be biased by allotting more random numbers to one direction than another. The network in Figure 10.1 demonstrates the technique. It was developed using the random number table at Appendix 2.

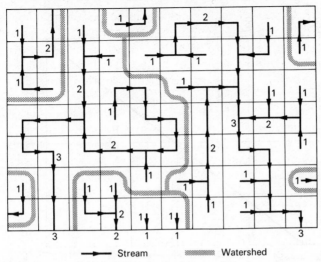

Figure 10.1. Networks simulated by using random numbers.

South was given twice the weighting of the other directions. Figures indicate the method of stream ordering, e.g. a stream below the confluence of two 1st-order streams becomes a 2nd-order stream; that below the confluence of two 2nd-order streams becomes a 3rd-order stream; and so on.

Ten digits were allotted as follows: west 0 and 1, north 2 and 3, east 4 and 5, south 6, 7, 8, and 9. The bias was to the south, representing a general tilt of the landscape in that direction. It will be seen that two major networks with clearly defined watersheds are beginning to develop, which, if classified as a drainage pattern using Strahler's method of stream numbering, would be the equivalent of third-order drainage basins.

A fifth-order drainage basin network generated by the random walk method is shown in Figure 10.2a (from Chorley and Haggett, 1967, after Leopold and Langbein, 1962). In this case the four directions were equally weighted, and the development of the main fifth-order 'river', in a general north to south direction, is the result of 'chance'. Figure 10.2b shows the number of 'streams' by stream orders plotted on semi-log graph paper. It will be seen that the ratio between the number of streams of one order and those of the next order, higher or lower, remains roughly constant; and therefore if plotted on semi-log paper the points will lie approximately on a straight line. This remains generally true of any random walk pattern of this type, provided the network is sufficiently large.

In this connection it is interesting to refer to Figure 6.3, which also shows a fifth-order drainage basin, but this time the network consists of the natural drainage pattern developed by the four major

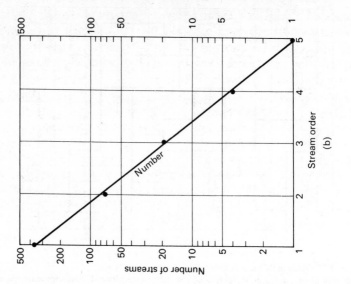

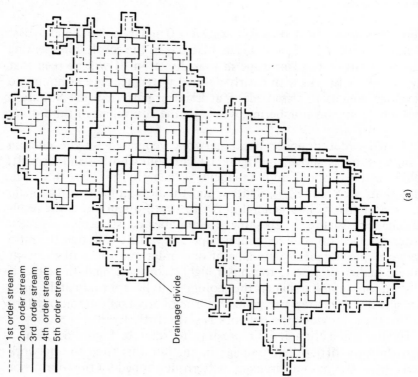

Figure 10.2. Drainage basin generated by random walk techniques (after Leopold & Langbein).

tributaries of the River Trent above Nottingham, together with the head waters of the Trent itself. Figure 10.3 shows the number of streams in each order plotted on semi-log graph paper. The reader will notice how very close these points lie to a straight line, and that therefore the *ratio* between the number of streams in successive orders must be nearly constant. (See p.90.) We have already calculated (Chapter 6.2) that there are differences, significant at the 0·05 level, *between* the drainage basins, in terms of the number of first-order streams. Now we see that, when considered *as a whole*, the network looks as though it approximates to one that has been randomly simulated. Of course, no conclusion can be drawn from a sample of 1, nor is it the purpose of this book to teach geomorphology. Nevertheless, it has been found in practice that real drainage

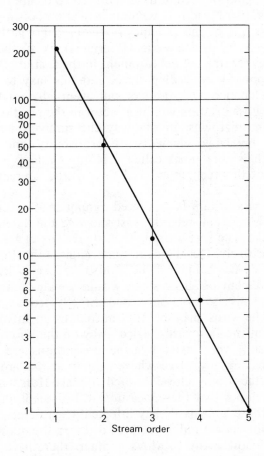

Figure 10.3. Stream order frequency for natural network shown in Fig. 6.3

patterns, if sufficiently large, often tend to resemble very closely those networks simulated by the random walk technique—indicating that a multiplicity of factors determines the development, the combined effect of which is approximately random.

10.3. SIMULATION MODELS—A DIFFUSION TECHNIQUE

The drainage network simulation described above is a model that has spatial aspects only—time is not involved. A basic assumption is that the natural drainage pattern must have sufficient time to develop to resemble the model. It also makes a *continuous* spatial pattern. That is, the network representing the flow of water is unbroken, unlike, for example, the colonization of new land, in which the settlements are separate entities, and therefore discontinuous.

A different kind of technique is required to create the model of a discontinuous distribution. It is frequently also necessary for *time* to be included as one of the variables of the model (e.g. the settlement pattern created by the colonizing farmers). A further important point is that every stage of development in the spatial pattern may be dependent upon the preceding stage, and this may to some extent depend upon chance. Our farmers will eventually establish villages and towns, whose location will be a factor in the positioning of more villages and more towns. In this way the simulation may be used predictively, involving elements of probability and 'chance', a method which is sometimes called a Monte Carlo type simulation. An example of this technique, used to construct a predictive model, is given below.

Over the years Britain has suffered a number of outbreaks of foot and mouth disease, and records exist showing the date and location of all outbreaks. The first warning that it exists in an area is when the disease is positively identified in some ailing animal, usually on a farm. Once this has happened the Ministry of Agriculture normally bans the movement of all livestock within a wide area of the initial outbreak. The disease may be passed directly from one animal to another, or the virus may be transmitted in an infected medium (e.g. mud from the farmyard carried away on the boots of a visitor). For this reason, very strict hygiene precautions are taken when entering or leaving any place where animals at risk are kept. (Foot and mouth attacks only cloven-hoofed animals. Horses are immune.) The Ministry hopes that these measures will succeed in confining the outbreak to those animals already infected when the disease was first recognized. But, as with all infectious diseases, the question arises as to where and how many previous contacts there have been. In the case of foot and mouth the source of infection may often be traced back to a cattle market. And the area in which the movement of cattle is prohibited in the first instance would normally include the farms and villages of every contact which might have visited the market. The initial pattern of outbreaks is therefore largely due to 'chance', depending on the farmers who happened to visit the market when the risk of infection was there; whether they had bought cattle, and brought them back to their farms; whether the cattle chanced to come in contact with infection while in the market; and so on.

One way of generating a random pattern of this kind is by the use of a grid drawn to a suitable scale and having numbered coordinates. In a real situation the number of actual initial outbreaks, and their location, could be used to predict the spread of the disease. The model may be used to simulate the pattern of a possible future occurrence by selecting a point of origin on the map and deciding arbitrarily on the number of initial outbreaks. These may then be located randomly, by placing the grid over the point of origin and using random numbers to determine the coordinates of each (Figure 10.4).

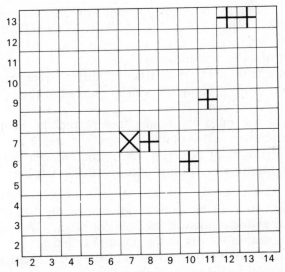

Figure 10.4. Grid with point of origin and five randomly located initial outbreaks

Once the movement of animals in an infected area has been prohibited and full hygiene precautions are in force a different situation exists. To understand how the infection is spread (and therefore to simulate this in the form of a model), it is necessary first to know something about the virus itself. The more important facts seem to be as follows:

1. The virus attacks only cloven-hoofed animals, such as cattle, sheep. pigs, etc. It is conveyed by contact between them, or with any material that has been infected, e.g. straw, feedstuffs, or farmyard dirt on boots or vehicle wheels.

2. It may be carried by the wind on particles of 2 to 10 microns (2 to 10 thousandths of one millimetre) in size.

3. Its life is prolonged by high relative humidity, by darkness, and by cold. It will survive freezing.

4. It is inhibited by dryness, and killed by sunlight.

5. Outside an animal's body it normally dies within about one week, though it may survive longer under favourable circumstances, and for up to at least one mouth in specially favourable conditions.

From the above it will be seen that weather must play a major part in the spread of the disease, and indeed all serious outbreaks have occurred during the winter months. The Meteorological Office conducted an investigation into the effect of winter conditions on outbreaks in Britain *prior* to the greatest of all epidemics in 1967. The following facts emerged:

1. Few outbreaks occurred upwind of an existing infection.

2. Outbreaks were restricted to $2\frac{1}{2}$ miles downwind in dry weather.

3. In wet weather outbreaks may take place further downwind, and furthest of all in conditions of light rain.

4. It was established that weather conditions were responsible for at least 80 per cent of occurrences of the spread of the infection, and maybe up to 95 per cent.

5. Wind is the most important factor, and the 80 per cent of outbreaks definitely attributable to weather are in a narrow zone downwind of the point of origin. *The prevailing wind is thus crucial.*

6. Of the outbreaks attributable to weather it has been established that:

33% occur within $1\frac{1}{4}$ miles of the point of origin
60% occur within 3 miles of the point of origin
75% occur within 6 miles of the point of origin
80% occur within 9 (approx.) miles of the point of origin
90% occur within $18\frac{1}{2}$ miles of the point of origin ⎫ Evidence less
95% occur within 31 miles of the point of origin ⎭ certain

It follows that once precautions are taken, spread of the virus is restricted (other than by wind) to a number of 'chance' factors. Wild animals, especially birds, may carry small particles of infected material to new areas. Hygiene regulations may very occasionally be imperfect and a farm visitor (e.g. a vet) has been known to bring infection to a 'clean' farm. The probability of these or other means of infection occurring may properly be treated as random, and may be taken to account for those cases not due to transport by wind.

The final factor involved in the spread of the disease (though not in its transmission) is the density of animals at risk in any given area. It is plain that where there is a relatively high density per acre of susceptible animals, the spread will be much quicker than in an area where the wind-borne virus will have a much smaller probability of coming in contact with a host in which it can multiply.

The way in which the disease spreads after precautions have been taken must be simulated bearing in mind the above factors. Of these wind appears to be by far the most important, but obviously allowance must be made for the residue of outbreaks not caused through transport by wind and attributable to 'chance' factors, like the flight of birds. One way of doing this is by using a probability matrix. The matrix shown in Figure 10.5 was constructed by allocating numbers between 000 and 529 to the squares of a regular grid. No numbers were allotted upwind of the point of origin as few outbreaks occur

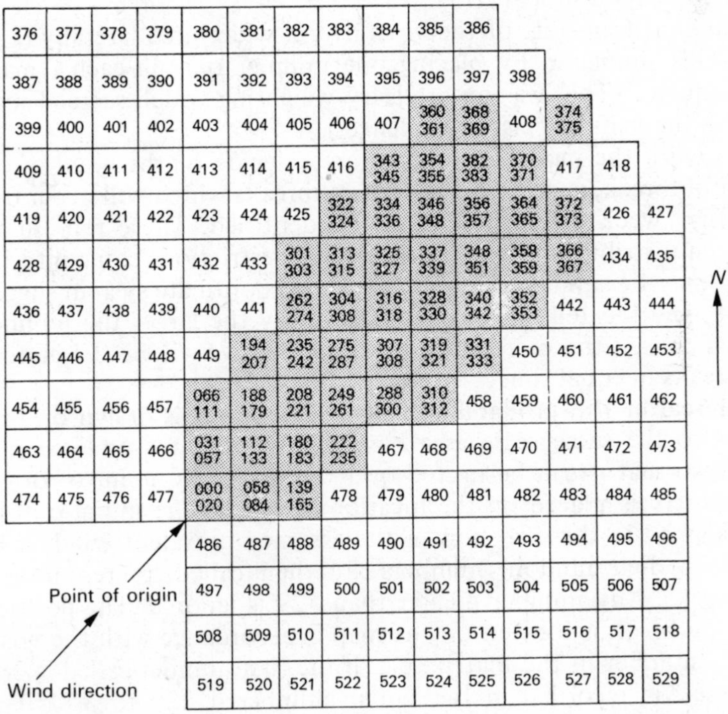

376	377	378	379	380	381	382	383	384	385	386					
387	388	389	390	391	392	393	394	395	396	397	398				
399	400	401	402	403	404	405	406	407	360 361	368 369	408	374 375			
409	410	411	412	413	414	415	416	343 345	354 355	382 383	370 371	417	418		
419	420	421	422	423	424	425	322 324	334 336	346 348	356 357	364 365	372 373	426	427	
428	429	430	431	432	433	301 303	313 315	325 327	337 339	348 351	358 359	366 367	434	435	
436	437	438	439	440	441	262 274	304 308	316 318	328 330	340 342	352 353	442	443	444	
445	446	447	448	449	194 207	235 242	275 287	307 308	319 321	331 333	450	451	452	453	
454	455	456	457	066 111	188 179	208 221	249 261	288 300	310 312	458	459	460	461	462	
463	464	465	466	031 057	112 133	180 183	222 235	467	468	469	470	471	472	473	
474	475	476	477	000 020	058 084	139 165	478	479	480	481	482	483	484	485	
				486	487	488	489	490	491	492	493	494	495	496	
				497	498	499	500	501	502	503	504	505	506	507	
				508	509	510	511	512	513	514	515	516	517	518	
				519	520	521	522	523	524	525	526	527	528	529	

Point of origin

Wind direction

N

Figure 10.5. Probability matrix associated with the spread of foot and mouth virus from a single point of origin

in this direction. The numbers are so arranged from a theoretical point of origin that they approximate to the percentages (given above) of outbreaks occurring downwind from a known point of infection. The percentage of the total of numbers used allotted to each square diminishes downwind with the established pattern of outbreaks outlined above. Thus the area represented by the square immediately adjacent *downwind* of the origin is at greatest risk; the probability of infection occurring there is over thirty times as great as in the area represented by the square immediately to the south; therefore thirty-one numbers (000 to 030) are allocated to the former, and only one (486) to the latter, whilst the numbers allotted to each square of the plume decrease proportionately to leeward until those at the furthermost point have only two. When random numbers are used to work the model in a simulated epidemic the probability of their location in the matrix depends on the number of numbers allotted to each square. The width of the plume representing downwind spread is arbitrary and a matter of judgement based on recorded cases. It represents minor local changes in wind direction over a short period.

Random transport of the virus over a wider area laterally (e.g. by birds), is simulated by placing one number only in each remaining grid square. (This was an arbitrary weighting which seemed reasonable in the light of existing evidence.)

To work the model, choose an area, decide on the point of origin of an outbreak, and the number of outbreaks which will occur during the first week after the disease is identified. These represent the random distribution of the virus *before* restrictions come into force. They may be simulated by means of the coordinate system suggested above, and located on the map. Obviously the larger the number of points chosen the more intensive *and* extensive will be the pattern of outbreaks in equal time.

Thereafter the probability matrix is used. It is drawn on tracing paper to the same scale as a base map of the area under consideration, so that it can be moved over the map. The point of origin of the matrix is placed on the location of one of the initial outbreaks and aligned in the direction of the assumed prevalent wind. A series of three digit random numbers are then called, representing new outbreaks. Any number greater than 529 is ignored. The position of each outbreak is located on the map in accordance with the position of the number in the matrix, e.g. if the first number called is 289 an outbreak is recorded in the square numbered 288 to 300. This is repeated for each initial outbreak. The pattern which appears on the map will be heavily weighted both in the direction of the wind chosen, and to the proximity of the point of origin, because this is the weighting given to the squares in the matrix. But chance locations will also appear unrelated to wind direction, when random numbers coincide with squares outside the wind plume. As it can be assumed that, once an outbreak at a specific place is identified, the Ministry policy of slaughter and disinfection will be effective, the matrix is centred only once on each location. The model is thus progressive in extending the pattern across the landscape.

No mention has been made so far of allowing for cattle density. To do this it is necessary to use a map on which built-up areas, moorland, and some approximation of cattle density are indicated. Clovenhoofed animals (except in zoos) do not live in towns. Any number falling on a built-up area, therefore, may be discounted. Similarly, as soon as the disease is identified; farmers in the area bring in any freerange animals they have on moorland, which may thus also be discounted. Giving a precise weighting to cattle density in the remaining parts of the map can only be an approximation and must remain a matter of individual judgement until further empirical evidence based on a detailed analysis of previous outbreaks, is available.

The method outlined above was used to simulate an epidemic of foot and mouth disease with a point of origin south-west of Oswestry, in the area where the first infected farm was identified in the great epidemic which began in October 1967. It was decided that five was a convenient number of initial chance outbreaks. Figure 10.4 shows them located on the grid. The selection of grid squares was random, but any chance selection of a square which fell on the high moorland to the west was discounted because of the very sparse animal population. The five points were then mapped, and the matrix (Figure 10.5) placed over each in turn, (including the origin). On every occasion five matrix squares were randomly selected and plotted on the map, to make 30 new locations. Each of these 30 new points was similarly treated, resulting in 150 further simulated outbreaks. During the real epidemic in 1967 the prevalent wind for the first three weeks was from the south-west, blowing strongly at times. The matrix was therefore orientated each time to simulate a south-west wind. To allow for the effect of cattle density (Figure 10.6) it was decided that in areas where there were more than 50 cattle per 100 acres of crops, grass, and rough grazing every outbreak should be counted, but if cattle density were lower than this (only a very small part of the area) two random numbers both falling in the same square would be required to constitute an outbreak.

The result of simulating 186 outbreaks of foot and mouth disease from a single point of origin by the method described is shown on

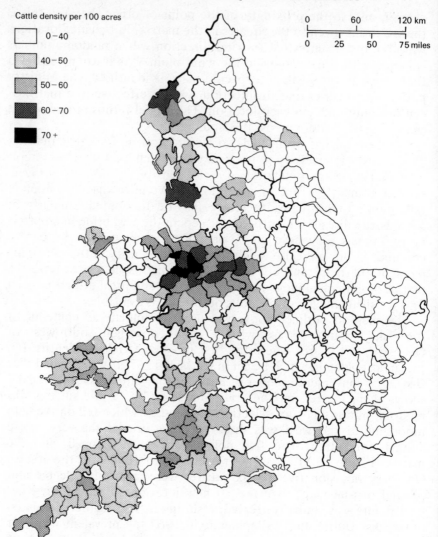

Cattle density per 100 acres

☐ 0–40

▨ 40–50

▨ 50–60

▧ 60–70

■ 70 +

0   60   120 km

0   25   50   75 miles

Figure 10.6. Cattle density per 100 acres crops, grass, and rough grazing in relation to N.A.A.S. Districts (1967)

Figure 10.7. The bias given to the spread by the effect of an assumed south-west wind is evident, although there is an expected peripheral scatter caused by chance. Figure 10.8 shows the actual location of real outbreaks which occurred during the first three weeks of the epidemic of 1967/8. The similarity of the areal spread of the disease is visually striking. It was tested more precisely, however, by superimposing a 3 X 5 grid on each map in turn. The number of outbreaks

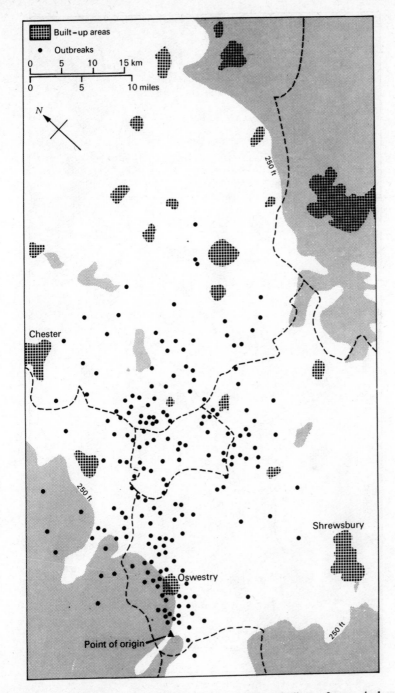

Figure 10.7. Simulation of 186 outbreaks of foot and mouth disease from a single point of origin

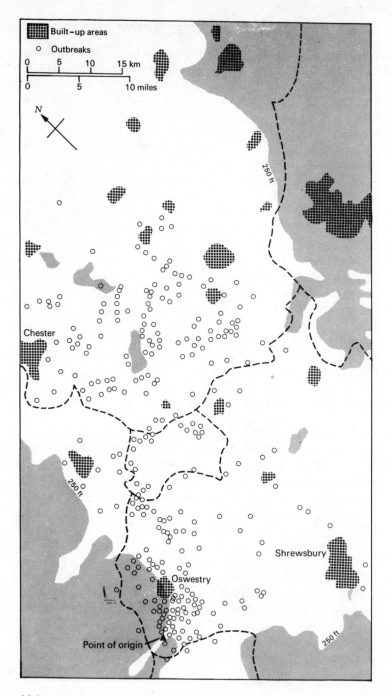

Figure 10.8. Actual outbreaks of foot and mouth disease in the first three weeks of the 1967/8 epidemic

located in each square, both simulated and real, was counted and recorded. The squares were then placed in rank order and Spearman's rank correlation coefficient calculated and found to be 0·67, which is significant at the 0·01 level. This shows that the model—based on information derived from previous epidemics—on this occasion is working reasonably well, even though the initial outbreaks were arbitrarily chosen to be 5, unrelated in number or precise location to the situation in 1967. Only the point of origin was the same. The model might therefore be used to approximate the kind of spread that might develop from an infection occurring at some future date.

The main differences are, first, that the model (which was not related to time) has not progressed as far as real epidemics in the first three weeks, and secondly, that the model presents a rather more symmetrical distribution than reality. The peripheral chance outbreaks are fairly evenly distributed in the model, whereas in reality there was a greater bias towards the east. This may be because insufficient allowance was made for the nature of the terrain to the west.

Perhaps by varying the weightings in the model, a closer approximation to reality may be achieved—and hence a more accurate insight into the relative strength of factors involved. Indeed, it is by such trial-and-error runs that simulation models are used to confirm, qualify, or reject our initial estimates.

10.4. THE GRAVITY MODEL

The family of gravity models are so called because they were based initially on a formula similar to the Newtonian law of gravitation. This stated that 'any two particles of matter attract one another with a force directly proportional to the product of their masses and inversely proportional to the square of the distance between them', and commonly expressed by the formula:

$$G \propto \frac{M_1 M_2}{D^2}$$

where G is the force of gravitation, M_1 and M_2 are the masses of two particles, and D is the distance between them. ($\propto$ is the sign meaning 'in proportion to'.)

The gravity models used by geographers mirror Newton's law in that the potential interaction between two places (generally towns or areas) depends upon some measurement of the 'size' of the places and some measurement of the 'distance' between them. In its simplest form potential interaction, e.g. between two towns, is assessed by multiplying the number of people resident in the one by those in the other and dividing by the distance between them, $I \propto \dfrac{P_1 P_2}{D}$. (Note that in this version D is not squared. The model is similar to, but not precisely the same as Newton's law.)

Interaction may be any aspect of dealings between the two places (visits by people, exchange of manufactured articles, and so on). The assumption is that the further the two places are apart the less interaction there will be, because time and cost increase with distance (the so called 'friction' of distance). Other assumptions associated with the model in this form are that the transport network is good and time and cost increase regularly in all directions.

Of course, the technique is extremely crude. Is the simple total population figure the best way to measure 'size'? (Mass in the Newtonian sense.) Would it not improve the model, as suggested by Haggett (1965, p.37), to multiply the population figure by the mean per capita income figure for the area (on the assumption that the wealthier the inhabitants were the greater the interaction was likely to be)? Alternatively, would size be more realistically represented by taking only the number of people *actively* employed? This might be argued on the grounds that retired people, children of school age, and the unemployed, contribute less to interaction with other places. Or would a combination of these be more realistic? Ultimately the decision rests with the *kind* of interaction the model is required to represent.

Similar problems arise over the question of distance, although in a small area like Britain where transport networks are good and the major centres are not very far apart, use of a simple distance figure may be adequate. (This has been used in the compilation of Figure 10.9.) But 'distance' does not necessarily have to be measured in spatial units. Time may be of much greater importance, as with some of the towns in the London commuting area. Distance measured in terms of cost might be an important limiting factor, especially when concerned with the exchange of heavy or bulky manufactures.

The type of transport used introduces its own complications. The distance factor between any two places is essentially different for goods carried by road than for those transported by rail. Or the introduction of a new type of service: for example, the opening of British Rail's Freightliner service between London and Edinburgh permitted fresh cakes baked in Leith to be sold as far away as London. Very different again is the effect of distance upon places with a coastal location when concerned with commodities most economically conveyed by sea.

Modifications to cater for all these (and many other) variations

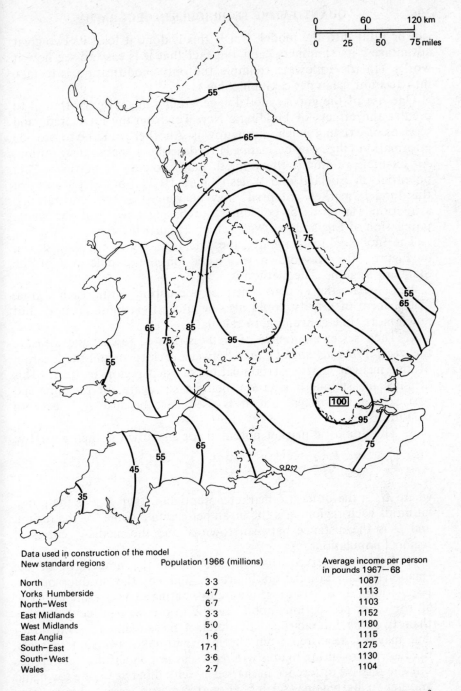

Data used in construction of the model

New standard regions	Population 1966 (millions)	Average income per person in pounds 1967—68
North	3·3	1087
Yorks Humberside	4·7	1113
North-West	6·7	1103
East Midlands	3·3	1152
West Midlands	5·0	1180
East Anglia	1·6	1115
South-East	17·1	1275
South-West	3·6	1130
Wales	2·7	1104

Figure 10.9. Domestic market potential in England and Wales expressed as a percentage of the maximum

can be built into the model. But if this is done it loses its two great advantages, its simplicity, and the fact that it is easy to see how it works. (In the following example the only modification is to take into account mean per capita income.)

One use of the gravity model in geography might be to attempt to predict the effects of building a New Town on the retail trade and services of existing neighbouring towns. Another could be to provide information (that can be mapped) to describe a spatial distribution. An example of this is the isopleth map showing domestic 'market potential' in England and Wales (Figure 10.9). For the purposes of this model 'market potential' may be viewed as the 'gravitational attraction' (in terms of consumer purchasing power) of the whole population of England and Wales at a particular location.

The model which generated the data from which the map shown in Figure 10.9 was drawn was based on the populations of the Standard Regions. The method is:

1. Calculate the median centre of population within each region. (The centre of gravity could also be used. It is more accurate but much more time-consuming to calculate.)

2. Then select a number of well-distributed towns, and measure the *straight line distance by land* between each town and the location of the median centre of population of each region in turn. (This distance measurement is crude, being a surrogate for the time and cost separating places, but it is used here because it is easy to measure.)

3. The domestic market potential for any town is then given by:

$$M_x = \sum \frac{P \cdot a}{D}$$

where M_x is the domestic market potential of town X, P is the population of each region, a is the mean per capita income for the region, and D is the distance between town X and the median centres of regional populations.

It will be observed that in this model the numerator of the fraction is one population figure, P, and not the product of two populations. This is because we are estimating the aggregate effect of all the other populations upon town X (a one-way influence) rather than two-way interaction. The units of measurement are arbitrary and may be measured from the map without conversion to scale, because the resulting figures will represent not an absolute difference but the relative difference in market potential between one town and another in its regional setting. (For this reason we are able to state $M_x = \Sigma (P \cdot a/D)$ rather than $M_x \propto \Sigma (P \cdot a/D)$.)

A difficulty arises in calculating the distance between a town and the centre of gravity of population in its home region, because in this case the town contributes to the regional centre of gravity, and may lie much closer to this point than to much of the regional population, e.g. in a peripheral distribution. It has been found that a reasonable compromise solution is to make the calculation of D for the home region as follows:

$$D = \tfrac{1}{2}(d_1 + d_2)$$

where D is the required distance, d_1 is the distance from the town to the median centre of population of the home region, and d_2 is the length of *half* the long axis of the region.

The result of the application of the model to selected towns in England and Wales is given in the form of an isopleth map in Figure 10.9. The result is an amalgam reflecting population and per capita income. It will be observed how greater market potential is reflected in the well-known 'axial belt' of population density, and in the higher income levels of the Midlands and the South-East. The assumptions on which the model is based are:

a. That 'market potential' is limited to England and Wales. (Exports are not counted.)
b. The market is proportional to the spending power of the population—hence it relates principally to consumer goods.
c. There is a uniform per capita pattern of consumption, varying only with income.
d. That distance decay in spatial interaction between places is proportional to the straight line distance separating them, rather than to some exponent of time- or cost-distance.

Classical (Weberian) industrial location theory has regarded the location and size of the market as 'given', and has tended to focus upon 'unit' profit. More recently the variation of market size with location has been recognized as affecting possible scale economies, and total (as opposed to unit) profits. This *type* of model could be used as a technique for measuring just that variation in market size with location, although more relevant variables and more precise data would need to be employed.

Another example of market potential, this time in south-east England, and expressed cartographically through the use of a gravity model, will be found in *Regional Studies*, Vol. 5, No. 4, Dec. 1971, p.251. And a most interesting series of maps expressing regional potential is contained in Vol. 3, No. 2, Sept. 1969, pp.202–8. These show, in terms of market potential, the calculated effects in Western Europe of the Treaty of Rome, the enlarged Common Market, containerization, and the projected Channel Tunnel.

This section began by referring to a family of gravity models, and illustrated these with the simplest case: $I \propto \dfrac{P_1 P_2}{D}$, which may also be expressed as $I = K \dfrac{P_1 P_2}{D}$, where K is a constant, partly determined by the units in which the variables are measured. Another member of the family is $I = K \dfrac{P_1 P_2}{D^2}$ which implies a greater friction of distance, since by, for instance trebling D, the value of I is reduced by a factor of nine instead of three. A whole branch of the family is represented by $I = K \cdot \dfrac{P_1 P_2}{D^m}$, where m is a power or exponent of D, determined by the friction of distance in a particular set of circumstances, and which might be expected to depend partly on the 'transferability' (ease of movement) of the people, goods, or information in terms of which interaction (I) is being estimated, and partly on the available means of transport or transfer. The way in which the most appropriate value of m can be ascertained in a particular case is illustrated by the example shown in Table 10.1 opposite.

We are interested in the number of working men commuting from Southend (a large dormitory town as well as a seaside resort) to neighbouring places of work. It is assumed that the number of commuters from Southend to its neighbours will conform to the equation $I = K \dfrac{P_1 P_2}{D^m}$, where P_1 is the size of the total number of in-commuters to each workplace (a more appropriate variable than population), P_2 is the total number of out-commuters from Southend, and D is the distance, in miles, between Southend and its neighbours. Since we are only interested in Southend commuters, P_2 is the same in every case, and can therefore be eliminated from the model, which therefore becomes $I = \dfrac{K \times P_1}{D^m}$. This can also be expressed as $\dfrac{I}{P_1} = \dfrac{K}{D^m}$ —i.e. the fraction of the commuting population of places of work which come from Southend is inversely proportional to D^m. If we take the log of both sides of the equation, we have $\log \dfrac{I}{P_1} = \log \dfrac{K}{D^m}$

$\therefore \quad \log \dfrac{I}{P_1} = \log K - m \log D.$

(For rules about manipulating the logs of quotients and powers, see pp.105/8.) Now, if K is a constant, so is $\log K$. If we call this c, we have:

$\log \dfrac{I}{P_1} = c - m \log D$, which is of the form $y = mx + c$, the equation

TABLE 10.1
Number of males commuting from Southend and elsewhere to selected parts of Essex and London (1971 – 10% sample)

	Basildon U.D.	Benfleet U.D.	Rayleigh U.D.	Thurrock U.D.	Rochford R.D.	City of London	Barking L.B.	Camden L.B.	Havering L.B.	Islington L.B.	Newham L.B.	Tower Hamlet L.B.	Westminster L.B.
I (Nos. commuting from Southend)	188	90	70	126	224	338	71	80	51	49	45	52	112
P_1 (Total nos. commuting in)	1218	261	233	1289	330	21877	3830	11971	1493	6725	3992	5331	28·485
D (Distance from Southend in miles)	11	7	6	16	3	35	24	37	20	36	29	32	38
$I \div P_1$	0·154	0·345	0·300	0·098	0·679	0·015	0·019	0·007	0·034	0·007	0·011	0·010	0·004

of a straight line, where y is $\log \dfrac{I}{P_1}$, x is $\log D$; m and c are constants determining the exact slope and position of the line. We can find values of m and c which give the best-fit regression equation for our data, using the method outlined on pp.256–8. (We first need to obtain values of x and y, i.e. of $\log D$ and $\log \dfrac{I}{P_1}$.)

The resulting values of m and c have been calculated to be $-2{\cdot}024$ and $1{\cdot}096$, so that the best-fit regression equation is $\log \dfrac{I}{P_1} = -2{\cdot}024$ $\log D + 1{\cdot}096$, and the member of the branch of gravity models which therefore best fits these commuting data is $I \propto \dfrac{P_1}{D^{2{\cdot}024}}$—i.e. the most appropriate exponent of D in this case is $2{\cdot}024$, which happens to be very close to the exponent in Newton's original law of gravity (2).

It is interesting to calculate *predicted* values of I from the model and compare them with actual values:

	Basildon	Benfleet	Rayleigh	Thurroch	Rochford	City	Barking	Camden	Havering	Islington	Newham	Tower Hamlet	Westminster
Predicted I	118	63	77	59	445	206	77	101	43	59	56	59	225
Actual I	118	90	70	126	224	338	71	80	51	49	45	52	112

Discrepancies may be due to:
   (i) Inappropriate selection of the variable to represent P_1—no account was taken of the *complementarity* of the population of Southend (in terms of age and social structure) and the varying types of employment available in neighbouring centres.
  (ii) Inappropriate selection of the variable to represent D—no account was taken of the variable *transferability* of the working population, which will depend on the efficiency of the road and rail connections between Southend and its neighbours.
 (iii) No account was taken of *intervening opportunities*—by no means all neighbouring workplaces were included in the data analysed, and Southend itself was omitted as a place of work.
 (iv) Random factors.

Nevertheless, there is a strong correlation ($r = 0.65$, significant at the 0·01 level) between the actual and predicted number of commuters from Southend. It can be said, therefore, that commuting numbers are in part related to gravitational 'forces', modified perhaps by complementarity, transferability, and intervening opportunities. To some extent these modifying factors would be accommodated by the model if more appropriate data were available.

Finally, the full family of gravity models is represented by the more general formula $I \propto \dfrac{P_1{}^k P_2{}^1}{D^m}$. Calculation of 'best-fit' values of k, l, and m requires techniques which are beyond the scope of this introductory text.

10.5. A MATHEMATICAL SLOPE MODEL

For many years the study of slope form and the processes related to it have occupied an important place in geomorphological research. Gilbert's suggestion (1909) that the upper convexity of hillslopes was related to the increasing volume of material transported downslope is an early example. More recently Young (1963) has devised a series of mathematical models to show development from an initial constant slope profile at an angle of 35 degrees, with a horizontal slope length of 1000 units, a height of 700 units, and level ground above. One of the useful features of these models (although not claimed by the author) is that his technique can be applied to any initial profile, and thus provides not only insight into the effect that different processes are likely to have on slope formation, but also a theoretical tool which may be used to predict the future development of some present slopes measured in the field. Sixteen slope models, together with details of calculation, are given in Young's paper. It is possible here to give only a simplified outline of the technique, and some selected models.

In constructing the models six factors are assumed likely to affect the subsequent development of the slope profile.

1. The slopes under consideration have no free face, are covered entirely by a regolith, and are in homogeneous material.

2. Slope form is affected by the transport of weathered material across it (in which soil creep is an important factor).

3. Slope form is also affected by direct removal (which in humid areas is normally the removal of dissolved material in solution by ground water).

4. The rate of transport will vary at any point with the sine of the slope angle, since this is the way in which gravitational pull varies when acting parallel to the ground surface.

5. If weathering takes place equally over the whole surface then, if slope retreat is to occur, the rate of transport of material may increase in proportion to the distance from the crest, since the material passing any point is proportional to the length of the slope above that point.

6. That there is unimpeded removal, but no downcutting, at the base of the slope. (Model 5 is an exception in which *no* basal removal is assumed.)

The assumptions contained in (1) above are common to all the models. Other factors are assumed singly or in combination, and are stated for each model.

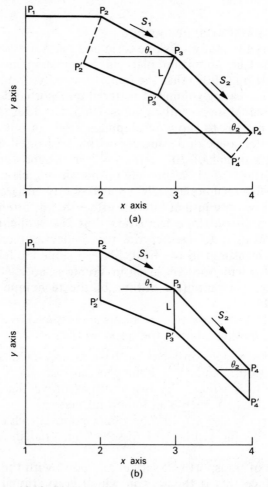

Figure 10.10. Diagrammatic representational of slopes (after Young)

Young adds this comment: '. . . it should be noted that the results of the slope models refer not necessarily to these specific processes but to any denudational processes which act in the ways stated in the assumptions.'

Figure 10.10a shows a theoretical slope represented diagrammatically. P_2, P_3, and P_4 are points along the profile 100 units apart measured horizontally. The profile lying between two points is assumed to be straight and is called a *section*.

Let us focus our attention on point P_3. θ_1 is the angle of slope of the section above P_3, and θ_2 the angle of slope of the section below P_3.

Because the models are concerned only with the slope *profile* we are able to treat the *volume* of material as an 'area' of soil transported past any point. If we call the volume of soil on the section above P_3 transported in one period of time S_1, and that on the section below S_2, then the gain or loss at P_3 will be proportional to $S_1 - S_2$. Figure 10.10a shows a loss of soil at P_3, with the new position of the point at P_3', and the distance between P_3 and P_3' denoted by L. It will be observed that the passage of each period of time would entail a different x-coordinate for each point affected. Fortunately this is unnecessary. Young shows that, by assuming ground loss takes place vertically (as in Figure 10.10b), the resulting inaccuracies are so slight that for practical purposes they can be ignored.

The method of calculation of slope model 1 (Figure 10.11) is given below. The assumptions associated with this model are: that removal is by downslope transportation only, varying with the sine of the slope angle; and that unimpeded base removal exists, but without downcutting.

It will be recalled that the initial slope from which the models are developed is assumed to have a horizontal length of 1000 units from P20 to P30, which is at a constant angle of 35. P20 is at a height of 700 units, and the ground from P18 to P20 assumed to be level. Points on the profile (shown in Figure 10.11) are at a horizontal distance of 100 units apart (x-coordinates) with a vertical interval of 70 similar units (y-coordinates) since tan $35° = 0.70$. Since the rate of movement downslope is taken to vary with the sine of the slope angle in this example, the amount of soil transported over one section (S) will be proportional to sin 35, i.e. 0.5736. This is an awkward number to work with and may conveniently be multiplied by a constant of 10. Therefore on every section of an initial slope soil transported downwards in one period of time is assumed to be:

$$S = k \sin \theta t$$

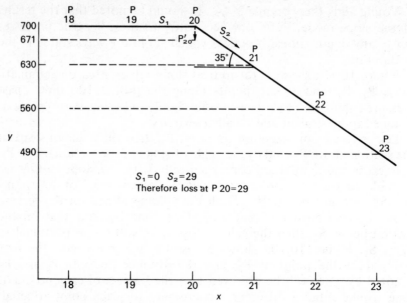

Figure 10.11. Part of the initial slope from which the models in Figure 10.13 were developed (after Young)

where S is the amount of soil moving downwards on one section, k is a constant of 10, θ is the slope angle, and t is time in arbitrary units.[1]

In Model 1 five times periods are assumed in each calculation, therefore the amount of soil moved downslope (S) on each section will be:

$$S = 10 \times 0\cdot5736 \times 5$$
$$= 29 \text{ (approx.)}.$$

The gain or loss at any point has been assumed as $S_1 - S_2$. During the first five time periods each point will lose and gain 29 units (because the initial angle of every section is assumed a constant 35), except P20, which loses 29 units without any corresponding gain. The new height of P20 (P20′) is therefore $700 - 29$, or 671.

Table 10.2 gives the calculation for part of Model 1. In this example unimpeded basal removal is assumed, but without any downcutting at the base of the slope, therefore point 30, marking the lowest point of the slope, remains unchanged. In Column 1 is the number of each point. Column 2_1 records the altitude of points of the *initial* slope. Above Point 20 the ground is flat at a height of 700. Below Point 20 the vertical distance between points is 70. The angle of each section is 35 degrees, shown in Column 3_1. Column 4_1 shows

TABLE 10.2

(1)	$(2)_1$	$(3)_1$	$(4)_1$	$(5)_1$	$(2)_2$	$(3)_2$	$(4)_2$	$(5)_2$	$(2)_3$	$(3)_3$	$(4)_3$	$(5)_3$	$(2)_4$	$(3)_4$	$(4)_4$	$(5)_4$	$(2)_5$	$(3)_5$	$(4)_5$	$(5)_5$	$(2)_6$	$(3)_6$
15	700	0°	0	0	700	0°	0	0	700	0°	0	0	700	0°	0	0	700	0°	0	0	700	0°
16	700	0°	0	−29	700	0°	0	−14	700	0°	0	−7	700	0°	0	−3	700	0°	0	−2	698	1°
17	700	0°	29	0	671	16°	14	−5	686	8°	7	−3	693	4°	3	−2	697	2°	2	−1	696	1°
18	700	0°	29	0	630	22°	19	−10	666	11°	10	−11	683	6°	5	−9	691	3°	3	−6	685	6°
19	700	0°	→		560	35°	29	0	620	25°	21	−5	655	16°	14	−5	674	10°	9	−2	672	7°
20	700	0°			490	35°	29	0	560	31°	26	−3	615	22°	19	−6	650	13°	11	−8	642	17°
21	630	35°				→			490	35°	29	0	557	30°	25	−3	609	22°	19	−5	604	21°
22	560	35°							420	35°	29	0	490	34°	28	−1	554	29°	24	−3	551	28°
23	490	35°								→			420	35°	29	0	489	33°	27	−2	487	33°
24	420	35°											350	35°	29	0	420	35°	29	0	420	34°
25	350	35°												→			350	35°	29	0	350	35°
26	280	35°																→			280	35°
27	210	35°																			210	35°
28	140	35°																			140	35°
29	70	35°																			70	35°
30	0	35°																			0	35°
t	5			—	5			—	5			—	5			—	5			—	—	
T	0				5				10				15				20				25	

(Note in group 2: "No ground loss")

Part of the calculations for Model 1

Col. 1 Number of point
Col. 2 Height of point
Col. 3 Angle of slope
Col. 4 Amount of soil transported
Col. 5 Gain or loss at each point
        For explanation see text.

t = time interval (in this case 5 periods)
T = total time elapsed

the amount of material, S, moving downslope on the section above each point in unit time. (In this case the time interval is assumed to be $5t$, and we have seen that the amount of material transported is approximately 29.)

Column 5_1 shows the ground loss at each point. In this case Point 20 has lost 29 units without any gain from above as it is the highest point on the slope. There is no change at other points downslope because loss is equal to gain.

Column 2_2 records the changes, if any, in the height of each point. In this case the only change is Point 20 which is reduced in height from 700 by 29 units to 671. Column 3_2 gives the new angles of slope formed as the result of the loss (in this case) of material. The method of calculation of these new angles may be seen using Figure 10.12. In the right-angled triangle ABC the new slope angle ACB (θ_1) has the natural tangent given by AB/BC. But BC is the horizontal distance between two points and is known to be 100 units, and AB has been calculated to be 29 units. We know therefore that θ_1 must be the angle having a tangent of 29/100. Or:

$$\theta_1 = \tan^{-1} \frac{29}{100}$$

$$= \tan^{-1} 0 \cdot 29$$

$$= 16° \text{ (approx.)}.$$

Similarly, we know the vertical interval between P20 and P21 is 70 units, and that AB is 29 units, so CD must be 41 units, therefore in the right-angled triangle CDE

$$\theta_2 = \tan^{-1} \frac{41}{100}$$

$$= \tan^{-1} 0 \cdot 41$$

$$= 22° \text{ (approx.)}.$$

Columns 4_2 and 5_2 are calculated as before, but using the new values from column 3_2. Thus the amount of soil gained (S_1) from the section above P20′ is $10 \times 0 \cdot 2756 \times 5 = 14$ approx. And the amount of soil lost (S_2) from the section below P20′ is $10 \times 0 \cdot 3746 \times 5 = 19$.

$$\text{Therefore gain or loss at P20}' = S_1 - S_2$$
$$= 14 - 19$$
$$= -5 \text{ units,}$$

and -5 is the new value in the row P20 for Column 5_2.

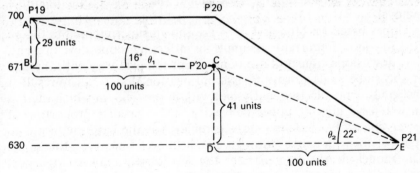

Figure 10.12.

The ground loss at each point in Column 5_2 is obtained from Column 4_2 by subtracting from the value given in each row the value in the row beneath it. Thus after 10 periods of time the ground loss for point 19 is $0 - 14 = -14$ units; for Point 20 is $+14 - 19 = -5$ units; and for Point 21 is $+19 - 29 = -10$ units.

Stages in the development of three slopes are given in Figure 10.13a, b, and c. All were devised from the same initial slope already decided. Model 1 assumed that the rate of transportation of material downslope varied with the sine of the slope angle ($S = (10 \sin \theta) \times 5$), and that no direct removal took place. The resulting curve is a gently rounded convexity with the steepest part tending toward the lower part of the slope. (For the purposes of this figure the short straight sections of the model have been reduced to smooth curves.) Such a form is found not infrequently in the field, but generally associated with an actively downcutting river.

Model 2 assumes all the conditions of Model 1 except that the rate of transportation varies not only with the sine of the angle of slope, but also in proportion to the distance from the crest. ($S = (10 \sin \theta) \times 5d$, where d is the distance in terms of the number of sections from the top of the slope.) It is interesting to see an early development of both convex and concave slope elements, together with parallel retreat of the constant (rectilinear) part of the slope. The curved profiles steadily extend downwards and upwards until they meet at a point of inflexion and the constant slope is consumed.

Model 3 assumes the initial conditions of Model 1, but denudation is by direct removal only, in which the rate of removal varies with the sine of the slope angle. In this case convex and concave elements occur at top and bottom, but a large part of the slope retreats in parallel at a constant angle. Young points out that: 'this is in agreement with the deductive system of W. Penk, who implicitly assumed a manner of action of processes which was in effect one of direct removal.'

Slopes based on Models 1 to 3 are shown superimposed in Figure 10.13d. Two important assumptions affecting slope development, that downslope transportation and direct removal vary with the sine of the slope angle, form the assumptions for Model 4. The result is the kind of profile frequently met with in the field, in which there is an extensive area of upper convexity, and a smaller proportion of lower concavity (because there is removal at the base), with a constant element retained throughout the development of the slope. The model clearly demonstrates the relationship between transport and direct removal. Young comments: 'the relative importance of parallel retreat and slope decline depends on the quantitative relationship between rates of transportation and (direct) removal. A constant feature, however, is that for any given set of assumptions, ground loss by direct removal becomes of relatively greater importance in the later stages of evolution.'

In Model 5 there is no basal removal. This condition is common when a slope terminates in the former flood plain of a river, or has at its lower end a valley terrace. Here transport downslope was taken to vary with the sine of the slope angle and distance from the crest. The resulting profile has a large concave element tapering gently downslope. Some field evidence suggests that progressively smaller particle size is to be expected as the angle of slope wanes towards its base. This model seems to accord well in this respect with the action of wash which is possibly the dominant process on this part of the slope. On the whole these models, generally in combination, seem to simulate the kind of slopes frequently found in the field in areas of temperate humid erosion, i.e., the simulation appears close to reality, thus indicating that the assumptions on which the models are based are probably valid.

In essence, this and the following model are no different from previous examples: variables are structured and weighted in accordance with theory; they are then fed with data, and the results compared with reality to test and perhaps improve the theory.

10.6. A COMPUTER-OPERATED SIMULATION MODEL

Many geomorphic processes can be simulated by the use of a computer. One of the main advantages gained in this way is that time can be speeded up so that the effect of process variables operating over many years can be observed within a few minutes. The model, in this case the computer programme, has first to be 'proved'. That is, it must be shown to reproduce with reasonable accuracy an

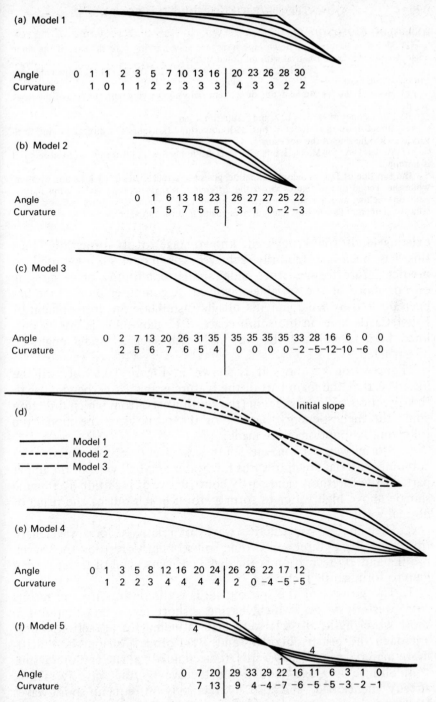

(a) Model 1

Angle	0	1	1	2	3	5	7	10	13	16	20	23	26	28	30
Curvature		1	0	1	1	2	2	3	3	3	4	3	3	2	2

(b) Model 2

Angle						0	1	6	13	18	23	26	27	27	25	22
Curvature							1	5	7	5	5	3	1	0	−2	−3

(c) Model 3

Angle	0	2	7	13	20	26	31	35	35	35	35	35	33	28	16	6	0	0
Curvature		2	5	6	7	6	5	4	0	0	0	0	−2	−5	−12	−10	−6	0

(d)

— Model 1
- - - - Model 2
– – – Model 3

Initial slope

(e) Model 4

Angle	0	1	3	5	8	12	16	20	24	26	26	22	17	12
Curvature		1	2	2	3	4	4	4	4	2	0	−4	−5	−5

(f) Model 5

1

4

4

1

Angle			0	7	20	29	33	29	22	16	11	6	3	1	0
Curvature				7	13	9	4	−4	−7	−6	−5	−5	−3	−2	−1

Figure 10.13. Five model slopes developed from the same initial slope, with level ground above

Assumptions for Figure 10.13

(a) Model 1. Removal by downslope transport only, varying with the sine of the slope angle. Unimpeded basal removal without downcutting.

(b) Model 2. As for Model 1, but transport varying with the sine of the slope angle *and* distance from crest.

(c) Model 3. As for Model 1 but direct removal only, varying with the sine of the slope angle.

(d) Final slopes of Models 1, 2, and 3 superimposed.

(e) Model 4. As for Model 1, but with downslope transport *and* direct removal, both varying with the sine of the slope angle.

(f) Model 5. As for Model 2, but with level ground below and this time *no* basal removal is assumed.

The top line of figures below each model gives the actual angles of each section of slope, while the second line of figures shows the *difference* in angle between each section and the next one below, and hence represents the amount of curvature. Negative values indicate concave elements. The vertical line indicates position of the initial crest.

existing landform through the known stages of its formation. Once this has been satisfactorily achieved the model may be used to predict future forms, either under similar conditions, or by altering one or more of the variables involved. A computer programme has been devised by King and McCullagh to simulate the development of Hurst Castle spit on the south coast of England, and is briefly outlined here to indicate the way in which such a model might be constructed.

The position of the spit is shown in Figure 10.14. It will be observed that the main part of the feature is aligned perpendicular to the direction of longest fetch (the south-west), from which direction come the highest-energy waves. For the purpose of the model the following assumptions were made.

1. Storm waves from the south-west are partly destructive by combing material down towards low water and below from the upper parts of the foreshore; and partly constructive at high tide by flinging shingle above high water to form a storm beach out of the reach of normal waves.

2. High-energy constructive waves of relatively long wavelength from the same direction would subsequently restore the beach profile, and, by refraction at the distal (seaward) end of the spit, may help to form gentle recurves.

3. The growth of the spit outwards is discontinuous, and results from westerly waves, which, having a short fetch, and therefore a short wavelength, are relatively little affected by refraction. These approach the beach obliquely and, by a process of longshore drift, move material rapidly eastwards, to accumulate at the end of the spit.

4. Shingle is moved round the recurve by the wave refraction already mentioned, and also by relatively infrequent short waves from the south and south-east. It is then built into a ridge by waves

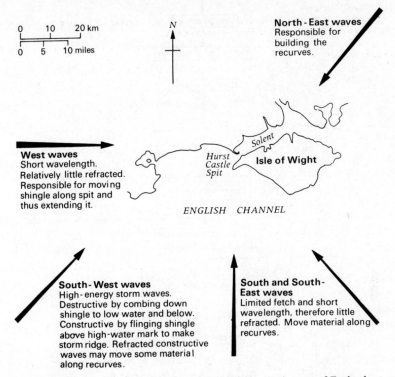

Figure 10.14. Location of Hurst Castle Spit on the south coast of England

approaching from the north-east down the Solent. The direction and frequencies of winds (and therefore of the waves generated by them) are thus obviously important in the construction of a model. (In south-west England wind frequencies are as follows: west winds 21 per cent, south-west winds 14 per cent (with 6 per cent force 5 or over), north-east winds 10 per cent, and south-east winds 6·5 per cent.)

Each recurve probably marks the position where the spit was stabilized for a period while shingle continued to be moved beyond the point, and before storm waves from the north-east were sufficiently large to form a new permanent recurve ridge. As the spit grew out from the land the water at the extremity grew deeper. Progressively larger amounts of shingle were required for each extension. There was a slowing down of the seaward movement of the point and an increasing period for each recurve to grow. With the passage of time, therefore, the recurves tend to grow longer.

All these factors were built into the computer programme, with weightings derived from observations. The computer reproduced the

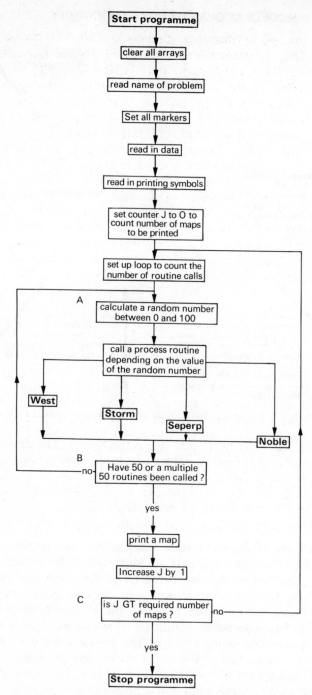

Figure 10.15. Flow-line diagram of computer programme.

outline of the spit by inserting numbers in a 50 X 60 matrix (Figure 10.16A). The programme was operated by four subroutines, (1) Storm, (2) West, (3) Seperp, and (4) Noble, each of which simulated the effect of the appropriate assumption listed above. The sub-routines shown in the programme flow diagram at Figure 10.15 are each allotted a series of numbers, and are activated in the programme by using random numbers. Thus 'Storm' might be allocated 61–70; 'West' 0–60; 'Seperp: (producing the vertical recurves) 71–80; and 'Noble' (producing the lateral recurves) 81–100. The weighting of each subroutine is a matter of judgement, based on wind frequency and direction, and is controlled by changing the range of values allotted to it. For example, with this distribution over a long period of runs subroutine 'West' would operate about 60 per cent of the time.

Incorporated with subroutine 'West' is a depth factor, which tends to slow down the growth of the spit eastwards under the influence of the westerly waves as the depth of water increases below 20 metres, and progressively larger amounts of material are required. Ultimately 'infinite' depth is assumed and outward building ceases.

The actual numbers printed in the matrix indicate the subroutine responsible for that part of the simulation. For example, 3s, 7s, and 9s are produced by 'Noble' and form the lateral recurves; 2s are produced by 'Storm'; and 1s by 'West'. An inspection of the simulated spit thus reveals the factors (the assumptions on which the subroutines were based) responsible for the formation of each of the parts. The programme is designed to print out a map (the numbers in the matrix) after every 50 calls of the subroutines. The result, after 4, 8, 16, and 25 runs of 50 is shown in Figure 10.16b. The closeness of the simulation to reality is an indication that the variables involved in the building of the spit were correctly identified and weighted.

To test the model a series of runs were made on the computer. In each case the same *values* were retained for the variables, but it is evident that differences in the *sequence* in which they are taken will have consequences in the development of the simulated spit. The sequence was therefore varied for each run by commencing with different initial random numbers. Figure 10.16a gives the result which most nearly resembles the real spit (shown in Figure 10.16c), and this becomes the model of the 'standard spit' used to extrapolate future development. In fact the similarity is quite remarkable. Of the stages in development shown in Figure 10.16b, 16 is closest to the actual spit. and 25 suggests a probable future stage.

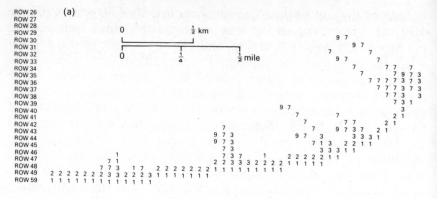

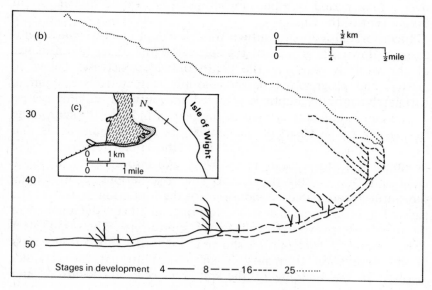

Figure 10.16. A simulation of Hurst Castle Spit

This model probably approximates as closely to reality as it is possible to get, and may be regarded as a very successful simulation. However, in considering the predictive reliability of this, as of all stochastic models, i.e. those which depend for their operation upon some element of chance, it is as well to bear in mind the authors' own warning:

One of the difficulties of the simulation model is . . . that the response of the feature to a given process variable is assumed to be known and to be always similar. This assumption may not be justified. The feature is certainly very drastically simplified in making the assumptions that are built into the model.

The criterion of success must rest on the degree to which the model can correctly simulate the form of the prototype . . . the main advantages of simulation models of this type are their ability to control the variables and their speed of operation. [But] like all models . . . they are very simplified replicas of reality. Their success depends on the degree to which the important variables are selected and simulated correctly.

10.7. LINEAR PROGRAMMING

All the other models described in this chapter can be called 'positive' or explanatory models, in that they seek to simulate, explain, or predict what is or will be rather than prescribe what ought to be. Linear programming is a technique which can be regarded as a normative or prescriptive model, because it determines the optimum allocation of resources under specific conditions. It is called linear programming because all the conditions are expressed in the form of linear relationships, analogous to linear equations but in fact called 'inequalities', which are then programmed with specified parameters. Generally, several variables (allocations) are involved, and the mathematical manipulation required for their solution is complex and tedious without the aid of a computer. But fortunately, the basic ideas underlying the model can be illustrated graphically by an artificially simple example, involving only two variables.

A farmer with 700 acres wishes to know what is the most profitable division of his land between cereals and sugar-beet, given the following conditions:

1. Standard labour requirements per acre of cereals and sugar-beet are 2 and 10 man-days respectively, and there are 10 men each working 300 man-days per annum, on the farm.

2. Other inputs (machinery, seed, chemicals, marketing, interest on borrowing) are estimated to be £50 and £100 per acre for cereals and sugar-beet respectively, and the farmer is able to raise £40 000.

3. Net profit per acre of cereals and sugar-beet are in the ratio of 1:4.

We begin by setting out the constraints as linear inequalities (so called because the two sides are not required to be equal). First, let the acreage under cereals be x_1

under sugar-beet be x_2

then:
(a) He can farm no more than 700 acres, therefore $x_1 + x_2 \leqslant 700$.
1st constraint

(b) Each acre of wheat consumes 2 man-days, each acre of beet consumes 10 man-days, out of an available total of 3000—

$$\therefore \quad 2x_1 + 10x_2 \leqslant 3000$$

or $x_1 + 5x_2 \leqslant 1500.$ 2nd constraint

(c) Each acre of wheat consumes £50 worth of other inputs, and each acre of sugar-beet £100, out of an available total of £40 000.

$$\therefore \quad 50x_1 + 100x_2 \leqslant 40\,000$$

or $x_1 + 2x_2 \leqslant 800.$ 3rd constraint

The other condition given is that net profit per acre on cereals and on sugar-beet are in the ratio of $1:4$. Therefore, to maximize total net profit (z), $x_1 + 4x_2$ must be maximized—i.e. $z = x_1 + 4x_2$ must be maximized. This equation expresses the objective of the operation, and is therefore called the *objective function*.

The three constraint inequalities and the objective function can now be graphed as in Figure 10.17a. Note the following:

1. Each inequality is bounded by a line representing the limiting equation—e.g. the limiting equation of $x_1 + x_2 \leqslant 700$ is $x_1 + x_2 = 700$. The inequality is therefore represented by a 'feasible' area below this line, since any point in this part of the graph satisfies the condition $x_1 + x_2 \leqslant 700$, and any point above it does not. (If the inequality were $x_1 + x_2 \geqslant 700$, it would be represented by the area above the line.)

2. x_1 and x_2 cannot have negative values. Therefore, the feasible areas represented by the inequalities are also bounded by the axes.

3. The polygon OPQRS represents that part of the graph which satisfies all three constraints, since it lies between the axes and the boundaries of all three inequalities. Values of x_1 and x_2 represented by any point in this polygon would therefore satisfy all three constraints. Any point outside the polygon would fail to satisfy at least one of the constraints, and would not therefore be feasible. The polygon is referred to as the 'total feasible area' or 'feasible region'.

4. The objective function, unlike the constraints, is not represented by an area on the graph, but by an infinite number of parallel lines, three of which are shown by dotted lines. They represent $z = x_1 + 4x_2$ for different values of z. This equation is usually reversed, as follows:

$$x_1 + 4x_2 = 800 \text{ (i.e. } z = 800) \text{ is the lowest dotted line;}$$
$$x_1 + 4x_2 = 1200 \text{ (i.e. } z = 1200) \text{ is the middle dotted line;}$$
$$x_1 + 4x_2 = 1600 \text{ (i.e. } z = 1600) \text{ is the highest dotted line.}$$

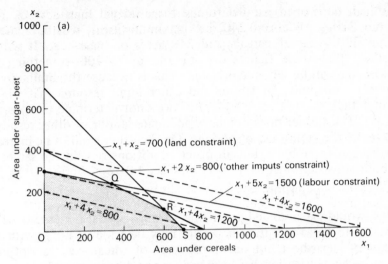

Figure 10.17a. Graphical solution to linear programming problem (see text)

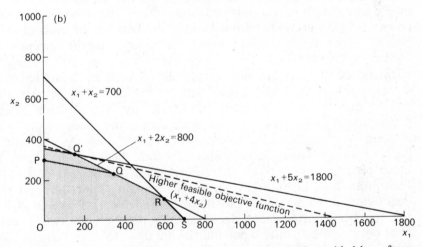

Figure 10.17b.   Graphical solution to linear programming problem with labour force increased by two

The further these lines are from the origin (O), the greater the value of z they represent.

Now the object is to maximize z while keeping within the feasible area defined by the constraints given. This is achieved at Q, one of the vertices of the total feasible area—the optimum is always to be found at one of the vertices. Here the objective function line furthest from the origin is just within the feasible area (it has not been drawn, but can be imaged parallel to the three dotted lines). At Q, x_1 is about 320, x_2 is about 240, i.e. maximum profit will be achieved from 320 acres of cereals and 240 acres of sugar-beet. It will be noticed that these figures do not add up to 700—something like 140 acres will be left uncultivated. This is because the limits are set by the availability of labour and other input resources (it will be noted that the line $x_1 + x_2 = 700$ does not determine the position of Q). The farmer does not have adequate labour or other resources to make the fullest use of his land. On the other hand, if he increases his labour force by 2, but all other conditions remain the same, the labour constraint inequality becomes $2x_1 + 10x_2 \leqslant 3600$

i.e. $x_1 + 5x_2 \leqslant 1800$.

This would be bounded by a line through Q' (see Figure 10.17b) parallel to PQ; the optimum would now be achieved at Q'; with 330 acres of sugar-beet but only 160 acres of wheat—i.e. the optimum area under cultivation has been *reduced* to 490 acres, and the area to be left uncultivated has been increased by the additional labour—surely an example of a counter-intuitive effect. The reason is that the farmer is being prevented from making the best use of his land by a lack of 'other inputs'—labour without adequate supporting capital is not fully effective. The above is, of course, a very unrealistic example. Farmers' constraints are not simply determined by input requirements, but also by demand (especially in the case of sugar-beet, for which quotas are allocated to farms by the British Sugar Corporation), while the weather, scale economies, keeping the land in good heart, maintaining an even workload through the year, and a host of other considerations also constrain farmers' decisions. If linear programming is to provide solutions to realistic problems, each constraint must be translated into an inequality function, and the goal (whether to maximize profits or minimize costs) into an objective function. Furthermore, decisions are rarely between two alternatives only: more than two possibilities are open to most farmers. But additional alternatives require additional variables in the inequalities and in the objective functions (one variable for each alternative), rendering a graphical solution impossible. Nevertheless, the principles and purpose of linear programming remain the same, however many the variables, and it is a useful tool in planning land-use and other forms of resource allocation, providing conditions can be specified accurately, and a computer is available to do the lengthy calculations required for a solution.

APPENDIX 1

Table of Probabilities Associated with Values of z in a Normal Distribution (correct to 3 decimal places)

z	Column A p	Column B p_1	Column C p_2
0·0	0·000	0·500	1·000
0·1	0·040	0·460	0·920
0·2	0·079	0·421	0·841
0·3	0·118	0·382	0·764
0·4	0·155	0·345	0·689
0·5	0·191	0·309	0·617
0·6	0·226	0·274	0·549
0·7	0·258	0·242	0·484
0·8	0·288	0·212	0·424
0·9	0·316	0·184	0·368
1·0	0·341	0·159	0·317
1·1	0·364	0·136	0·271
1·2	0·385	0·115	0·230
1·3	0·403	0·097	0·193
1·4	0·419	0·081	0·162
1·5	0·433	0·067	0·134
1·6	0·445	0·055	0·110
1·7	0·455	0·045	0·089
1·8	0·464	0·036	0·072
1·9	0·471	0·029	0·057
*1·96	0·475	0·025	0·050
2·0	0·477	0·023	0·046
2·1	0·482	0·018	0·036
2·2	0·486	0·014	0·028
2·3	0·489	0·011	0·021
2·4	0·492	0·008	0·016
2·5	0·494	0·006	0·012
*2·58	0·495	0·005	0·010
2·6	0·495	0·005	0·009
2·7	0·496	0·004	0·007
2·8	0·497	0·003	0·005

z	Column A p	Column B p_1	Column C p_2
2·9	0·498	0·002	0·004
3·0	0·499	0·001	0·003
3·1	0·499	0·001	0·002
3·2	0·499	0·001	0·001
3·3	0·499	0·001	0·001
3·4	0·500	0·000	0·001
3·5	0·500	0·000	0·000

Note:

Col. A: p = the probability of a value lying between the axis of symmetry (i.e. $z = 0$) and given value of z (a one-tailed probability).

Col. B: p_1 = the probability of a value being more extreme than z (a one-tailed probability).

Col. C: p_2 = the probability of a value being more extreme than either $+z$ or $-z$ (a two-tailed probability).

*Critical values of z corresponding to the 0·05 and 0·01 levels (two-tailed) have been given to two decimal places.

After Lindley and Miller (1953).

APPENDIX 2

Random Sampling Numbers

20 17	42 28	23 17	59 66	38 61	02 10	86 10	51 55	92 52
74 49	04 49	03 04	10 33	53 70	11 54	48 63	94 60	94 49
94 70	49 31	38 67	23 42	29 65	40 88	78 71	37 18	48 64
22 15	78 15	69 84	32 52	32 54	15 12	54 02	01 37	38 37
93 29	12 18	27 30	30 55	91 87	50 57	58 51	49 36	12 53
45 04	77 97	36 14	99 45	52 95	69 85	03 83	51 87	85 56
44 91	99 49	89 39	94 60	48 49	06 77	64 72	59 26	08 51
16 23	91 02	19 96	47 59	89 65	27 84	30 92	63 37	26 24
04 50	65 04	65 65	82 42	70 51	55 04	61 47	88 83	99 34
32 70	17 72	03 61	66 26	24 71	22 77	88 33	17 78	08 92
03 64	59 07	42 95	81 39	06 41	20 81	92 34	51 90	39 08
62 49	00 90	67 86	83 48	31 83	19 07	67 68	49 03	27 47
61 00	95 86	98 36	14 03	48 88	51 07	33 40	06 86	33 76
89 03	90 49	28 74	21 04	09 96	60 45	22 03	52 80	01 79
01 72	33 85	52 40	60 07	06 71	89 27	14 29	55 24	85 79
27 56	49 79	34 34	32 22	60 53	91 17	33 26	44 70	93 14
49 05	74 48	10 55	35 25	24 28	20 22	35 66	66 34	26 35
49 74	37 25	97 26	33 94	42 23	01 28	59 58	92 69	03 66
20 26	22 43	88 08	19 85	08 12	47 65	65 63	56 07	97 85
48 87	77 96	43 39	76 93	08 79	22 18	54 55	93 75	97 26
08 72	87 46	75 73	00 11	27 07	05 20	30 85	22 21	04 67
95 97	98 62	17 27	31 42	64 71	46 22	32 75	19 32	20 99
37 99	57 31	70 40	46 55	46 12	24 32	36 74	69 20	72 10
05 79	58 37	85 33	75 18	88 71	23 44	54 28	00 48	96 23
55 85	63 42	00 79	91 22	29 01	41 39	51 50	36 65	26 11
67 28	96 25	68 36	24 72	03 85	49 24	05 69	64 86	08 19
85 86	94 78	32 59	51 82	86 43	73 84	45 60	89 57	06 87
40 10	60 09	05 88	78 44	63 13	58 25	37 11	18 47	75 62
94 55	89 48	90 80	77 80	26 89	87 44	23 74	66 20	20 19
11 63	77 77	23 20	33 62	62 19	29 03	94 15	56 37	14 09
64 00	26 04	54 55	38 57	94 62	68 40	26 04	24 25	03 61
50 94	13 23	78 41	60 58	10 60	88 46	30 21	45 98	70 96
66 98	37 96	44 13	45 05	34 59	75 85	48 97	27 19	17 85
66 91	42 83	60 77	90 91	60 90	79 62	57 66	72 28	08 70
33 58	12 18	02 07	19 40	21 29	39 45	90 42	58 84	85 43
52 49	70 16	72 40	73 05	50 90	02 04	98 24	05 30	27 25
74 98	93 99	78 30	79 47	96 62	45 58	40 37	89 76	84 41
50 26	54 30	01 88	69 57	54 45	69 88	23 21	05 69	93 44
49 46	61 89	33 79	96 84	28 34	19 35	28 73	39 59	56 34
19 64	13 44	78 39	73 88	62 03	36 00	25 96	86 76	67 90
64 17	47 67	87 59	81 40	72 61	14 00	28 28	55 86	23 38
18 43	97 37	68 97	56 56	57 95	01 88	11 89	48 07	42 07
65 58	60 87	51 09	96 61	15 53	66 81	66 88	44 75	37 01
79 90	31 00	91 14	85 65	31 75	43 15	45 93	64 78	34 53
07 23	00 15	59 05	16 09	94 42	20 40	63 76	65 67	34 11
90 98	14 24	01 51	95 46	30 32	33 19	00 14	19 28	40 51
53 82	62 02	21 82	34 13	41 03	12 85	65 30	00 97	56 30
98 17	26 15	04 50	76 25	20 33	54 84	39 31	23 33	59 64
08 91	12 44	82 40	30 62	45 50	64 54	65 17	89 25	59 44
37 21	46 77	84 87	67 39	85 54	97 37	33 41	11 74	90 50

After Lindley and Miller (1953).

APPENDIX 3

Student's t-Distribution

Table of values of t corresponding to specified *two-tailed* probabilities and degrees of freedom

Degrees of Freedom	$p = 0.1$ $p' = 90\%$	$p = 0.05$ $p' = 95\%$	$p = 0.02$ $p' = 98\%$	$p = 0.01$ $p' = 99\%$	$p = 0.001$ $p' = 99.9\%$
1	6·31	12·71	31·82	63·66	636·62
2	2·92	4·30	6·97	9·93	31·60
3	2·35	3·18	4·54	5·84	12·94
4	2·13	2·78	3·75	4·60	8·61
5	2·02	2·57	3·37	4·03	6·86
6	1·94	2·45	3·14	3·71	5·96
7	1·90	2·37	3·00	3·50	5·41
8	1·86	2·31	2·90	3·36	5·04
9	1·83	2·26	2·82	3·25	4·78
10	1·81	2·23	2·76	3·17	4·59
11	1·80	2·20	2·72	3·11	4·44
12	1·78	2·18	2·68	3·06	4·32
13	1·77	2·16	2·65	3·01	4·22
14	1·76	2·15	2·62	2·98	4·14
15	1·75	2·13	2·60	2·95	4·07
16	1·75	2·12	2·58	2·92	4·02
17	1·74	2·11	2·57	2·90	3·97
18	1·73	2·10	2·55	2·88	3·92
19	1·73	2·09	2·54	2·86	3·88
20	1·73	2·09	2·53	2·85	3·85
21	1·72	2·08	2·52	2·83	3·82
22	1·72	2·07	2·51	2·82	3·79
23	1·71	2·07	2·50	2·81	3·77
24	1·71	2·06	2·49	2·80	3·75
25	1·71	2·06	2·49	2·79	3·73
26	1·71	2·06	2·48	2·78	3·71
27	1·70	2·05	2·47	2·77	3·69
28	1·70	2·05	2·47	2·76	3·67
29	1·70	2·05	2·46	2·76	3·66
30	1·70	2·04	2·46	2·75	3·65
40	1·68	2·02	2·42	2·70	3·55
60	1·67	2·00	2·39	2·66	3·46

Note:

p is the two-tailed probability of a value being more extreme than t.

p' is the two-tailed probability (expressed as a percentage) of a value being less extreme than t.

From Alder and Roessler (1964); after Fisher and Yates (1953).

APPENDIX 4

The Chi Square Test

The critical values of chi square given below show the probability that the calculated value of χ^2 is the result of a chance distribution. The *larger* the value of χ^2 the *smaller* is the probability that H_0 is correct.

df	0·10	0·05	0·01	0·001
1	2·71	3·84	6·64	10·83
2	4·60	5·99	9·21	13·82
3	6·25	7·82	11·34	16·27
4	7·78	9·49	13·28	18·46
5	9·24	11·07	15·09	20·52
6	10·64	12·59	16·81	22·46
7	12·02	14·07	18·48	24·32
8	13·36	15·51	20·09	26·12
9	14·68	16·92	21·67	27·88
10	15·99	18·31	23·21	29·59
11	17·28	19·68	24·72	31·26
12	18·55	21·03	26·22	32·91
13	19·81	22·36	27·69	34·53
14	21·06	23·68	29·14	36·12
15	22·31	25·00	30·58	37·70
16	23·54	26·30	32·00	39·29
17	24·77	27·59	33·41	40·75
18	25·99	28·87	34·80	42·31
19	27·20	30·14	36·19	43·82
20	28·41	31·41	37·57	45·32
21	29·62	32·67	38·93	46·80
22	30·81	33·92	40·29	48·27
23	32·01	35·17	41·64	49·73
24	33·20	36·42	42·98	51·18
25	34·38	37·65	44·31	52·62
26	35·56	38·88	45·64	54·05
27	36·74	40·11	46·96	55·48
28	37·92	41·34	48·28	56·89
29	39·09	42·56	49·59	58·30
30	40·26	43·77	50·89	59·70

From Siegel (1956); after Fisher and Yates (1953).

APPENDIX 5

The F-distribution for Variance Ratio Test

The numbers in the table are the critical values of F (the variance ratio) for the 0·05 level of significance. If the calculated value of F is *less* than the critical value, the difference in sample variances may be regarded as being due to chance.

		3	4	5	6	8	10	15	20	30	40
		\multicolumn{10}{c}{*Degrees of freedom of larger estimate of variance*}
Degrees of freedom of smaller estimate of variance	3	9·3	9·1	9·0	8·9	8·9	8·8	8·7	8·7	8·6	8·6
	4	6·6	6·4	6·3	6·2	6·1	6·0	5·9	5·8	5·8	5·7
	5	5·4	5·2	5·1	5·0	4·8	4·7	4·6	4·6	4·5	4·5
	6	4·8	4·5	4·4	4·4	4·2	4·1	3·9	3·9	3·8	3·8
	8	4·1	3·8	3·7	3·6	3·4	3·4	3·2	3·2	3·1	3·0
	10	3·7	3·5	3·3	3·2	3·1	3·0	2·9	2·8	2·7	2·7
	15	3·3	3·1	2·9	2·8	2·6	2·5	2·4	2·3	2·3	2·2
	20	3·1	2·9	2·7	2·6	2·5	2·4	2·2	2·1	2·0	2·0
	30	2·9	2·7	2·5	2·4	2·3	2·2	2·0	1·9	1·8	1·8
	40	2·8	2·6	2·5	2·3	2·2	2·1	1·9	1·8	1·7	1·7

From Alder and Roessler (1964); after Merrington and Thompson (1943).

APPENDIX 6

The Mann-Whitney U *Test*

For Tables A to F: The probability read off in the table must be *less than* the rejection level (α) if U is to be regarded as significant.

For Tables G to J: The calculated value of U must be *less than* the critical value at the specified rejection level if U is to be regarded as significant.

Note also: The probability values in Tables A to F are one-tailed. For one-tailed values, simply halve the probabilities.

A
$n_2 = 3$

U \ n_1	1	2	3
0	0·250	0·100	0·050
1	0·500	0·200	0 0·100

B
$n_2 = 4$

U \ n_1	1	2	3	4
0	0·200	0·067	0·028	0·014
1	0·400	0·133	0·057	0·029
2	0·600	0·267	0·114	0·057
3		0·400	0·200	0·100

C
$n_2 = 5$

U \ n_1	1	2	3	4	5
0	0·167	0·047	0·018	0·008	0·004
1	0·333	0·095	0·036	0·016	0·008
2	0·500	0·190	0·071	0·032	0·016
3	0·667	0·286	0·125	0·056	0·028
4		0·429	0·196	0·095	0·048
5		0·571	0·286	0·143	0·075

Appendix 6 (Contd.)

D
$n_2 = 6$

U \ n_1	1	2	3	4	5	6
0	0·143	0·036	0·012	0·005	0·002	0·001
1	0·286	0·071	0·024	0·010	0·004	0·002
2	0·428	0·143	0·048	0·019	0·009	0·004
3	0·571	0·214	0·083	0·033	0·015	0·008
4		0·321	0·131	0·057	0·026	0·013
5		0·429	0·190	0·086	0·041	0·021
6		0·571	0·274	0·129	0·063	0·032
7			0·357	0·176	0·089	0·047
8			0·452	0·238	0·123	0·066
9			0·548	0·305	0·165	0·090

E
$n_2 = 7$

U \ n_1	1	2	3	4	5	6	7
0	0·125	0·028	0·008	0·003	0·001	0·001	0·000
1	0·250	0·056	0·017	0·006	0·003	0·001	0·001
2	0·375	0·111	0·033	0·012	0·005	0·002	0·001
3	0·500	0·167	0·058	0·021	0·009	0·004	0·002
4	0·625	0·250	0·092	0·036	0·015	0·007 −	0·003
5		0·333	0·133	0·055	0·024	0·011	0·006
6		0·444	0·192	0·082	0·037	0·017	0·009
7		0·556	0·258	0·115	0·053	0·026	0·013
8			0·333	0·158	0·074	0·037	0·019
9			0·417	0·206	0·101	0·051	0·027
10			0·500	0·264	0·134	0·069	0·036
11			0·583	0·324	0·172	0·090	0·049
12				0·394	0·216	0·117	0·064
13				0·464	0·265	0·147	0·082

Appendix 6 (Contd.)

F
$n_2 = 8$

n_1 / U	1	2	3	4	5	6	7	8
0	0·111	0·022	0·006	0·002	0·001	0·000	0·000	0·000
1	0·222	0·044	0·012	0·004	0·002	0·001	0·000	0·000
2	0·333	0·089	0·024	0·008	0·003	0·001	0·001	0·000
3	0·444	0·133	0·042	0·014	0·005	0·002	0·001	0·001
4	0·556	0·200	0·067	0·024	0·009	0·004	0·002	0·001
5		0·267	0·097	0·036	0·015	0·006	0·003	0·001
6		0·356	0·139	0·055	0·023	0·010	0·005	0·002
7		0·444	0·188	0·077	0·033	0·015	0·007	0·003
8		0·556	0·248	0·107	0·047	0·021	0·010	0·005
9			0·315	0·141	0·064	0·030	0·014	0·007
10			0·387	0·184	0·085	0·041	0·020	0·010
11			0·461	0·230	0·111	0·054	0·027	0·014
12			0·539	0·285	0·142	0·071	0·036	0·019
13				0·341	0·177	0·091	0·047	0·025
14				0·404	0·217	0·114	0·060	0·032
15				0·467	0·262	0·141	0·076	0·041
16				0·533	0·311	0·172	0·095	0·052
17					0·362	0·207	0·116	0·065
18					0·416	0·245	0·140	0·080
19					0·472	0·286	0·168	0·097

From Siegal (1956); after Mann and Whitney (1947).

Appendix 6 (Contd.)

G
Critical Values of U for a two-tailed test at $\alpha = 0.002$ and a one-tailed test at $\alpha = 0.001$

n_1 \ n_2	9	10	11	12	13	14	15	16	17	18	19	20
1												
2												
3									0	0	0	0
4		0	0	0	1	1	1	2	2	3	3	3
5	1	1	2	2	3	3	4	5	5	6	7	7
6	2	3	4	4	5	6	7	8	9	10	11	12
7	3	5	6	7	8	9	10	11	13	14	15	16
8	5	6	8	9	11	12	14	15	17	18	20	21
9	7	8	10	12	14	15	17	19	21	23	25	26
10	8	10	12	14	17	19	21	23	25	27	29	32
11	10	12	15	17	20	22	24	27	29	32	34	37
12	12	14	17	20	23	25	28	31	34	37	40	42
13	14	17	20	23	26	29	32	35	38	42	45	48
14	15	19	22	25	29	32	36	39	43	46	50	54
15	17	21	24	28	32	36	40	43	47	51	55	59
16	19	23	27	31	35	39	43	48	52	56	60	65
17	21	25	29	34	38	43	47	52	57	61	66	70
18	23	27	32	37	42	46	51	56	61	66	71	76
19	25	29	34	40	45	50	55	60	66	71	77	82
20	26	32	37	42	48	54	59	65	70	76	82	88

Appendix 6 (Contd.)

H
Critical values of U for a two-tailed test at $\alpha = 0.02$
and a one-tailed test at $\alpha = 0.01$

n_1 \ n_2	9	10	11	12	13	14	15	16	17	18	19	20
1												
2					0	0	0	0	0	0	1	1
3	1	1	1	2	2	2	3	3	4	4	4	5
4	3	3	4	5	5	6	7	7	8	9	9	10
5	5	6	7	8	9	10	11	12	13	14	15	16
6	7	8	9	11	12	13	15	16	18	19	20	22
7	9	11	12	14	16	17	19	21	23	24	26	28
8	11	13	15	17	20	22	24	26	28	30	32	34
9	14	16	18	21	23	26	28	31	33	36	38	40
10	16	19	22	24	27	30	33	36	38	41	44	47
11	18	22	25	28	31	34	37	41	44	47	50	53
12	21	24	28	31	35	38	42	46	49	53	56	60
13	23	27	31	35	39	43	47	51	55	59	63	67
14	26	30	34	38	43	47	51	56	60	65	69	73
15	28	33	37	42	47	51	56	61	66	70	75	80
16	31	36	41	46	51	56	61	66	71	76	82	87
17	33	38	44	49	55	60	66	71	77	82	88	93
18	36	41	47	53	59	65	70	76	82	88	94	100
19	38	44	50	56	63	69	75	82	88	94	101	107
20	40	47	53	60	67	73	80	87	93	100	107	114

Appendix 6 (Contd.)

I

Critical Values of U for a two-tailed test at $\alpha = 0.05$
and a one-tailed test at $\alpha = 0.025$

n_1 \\ n_2	9	10	11	12	13	14	15	16	17	18	19	20
1												
2	0	0	0	1	1	1	1	1	2	2	2	2
3	2	3	3	4	4	5	5	6	6	7	7	8
4	4	5	6	7	8	9	10	11	11	12	13	13
5	7	8	9	11	12	13	14	15	17	18	19	20
6	10	11	13	14	16	17	19	21	22	24	25	27
7	12	14	16	18	20	22	24	26	28	30	32	34
8	15	17	19	22	24	26	29	31	34	36	38	41
9	17	20	23	26	28	31	34	37	39	42	45	48
10	20	23	26	29	33	36	39	42	45	48	52	55
11	23	26	30	33	37	40	44	47	51	55	58	62
12	26	29	33	37	41	45	49	53	57	61	65	69
13	28	33	37	41	45	50	54	59	63	67	72	76
14	31	36	40	45	50	55	59	64	67	74	78	83
15	34	39	44	49	54	59	64	70	75	80	85	90
16	37	42	47	53	59	64	70	75	81	86	92	98
17	39	45	51	57	63	67	75	81	87	93	99	105
18	42	48	55	61	67	74	80	86	93	99	106	112
19	45	52	58	65	72	78	85	92	99	106	113	119
20	48	55	62	69	76	83	90	98	105	112	119	127

From Siegal (1956); after Auble (1953).

Appendix 6 (Contd.)

J
Critical Values of U for a two-tailed test at $\alpha = 0.10$ and a one-tailed test at $\alpha = 0.05$

n_1 \ n_2	9	10	11	12	13	14	15	16	17	18	19	20
1											0	0
2	1	1	1	2	2	2	3	3	3	4	4	4
3	3	4	5	5	6	7	7	8	9	9	10	11
4	6	7	8	9	10	11	12	14	15	16	17	18
5	9	11	12	13	15	16	18	19	20	22	23	25
6	12	14	16	17	19	21	23	25	26	28	30	32
7	15	17	19	21	24	26	28	30	33	35	37	39
8	18	20	23	26	28	31	33	36	39	41	44	47
9	21	24	27	30	33	36	39	42	45	48	51	54
10	24	27	31	34	37	41	44	48	51	55	58	62
11	27	31	34	38	42	46	50	54	57	61	65	69
12	30	34	38	42	47	51	55	60	64	68	72	77
13	33	37	42	47	51	56	61	65	70	75	80	84
14	36	41	46	51	56	61	66	71	77	82	87	92
15	39	44	50	55	61	66	72	77	83	88	94	100
16	42	48	54	60	65	71	77	83	89	95	101	107
17	45	51	57	64	70	77	83	89	96	102	109	115
18	48	55	61	68	75	82	88	95	102	109	116	123
19	51	58	65	72	80	87	94	101	109	116	123	130
20	54	62	69	77	84	92	100	107	115	123	130	138

From Siegal (1956); after Auble (1953).

APPENDIX 7

Table of Critical Values of T in the Wilcoxon Matched-Pairs Signed-Ranks Test.
T must be less than critical value to be significant at specified level.

	Level of significance for one-tailed test		
	0·025	0·01	0·005
$N =$ no. of pairs	Level of significance for two-tailed test		
	0·05	0·02	0·01
6	0	—	—
7	2	0	—
8	4	2	0
9	6	3	2
10	8	5	3
11	11	7	5
12	14	10	7
13	17	13	10
14	21	16	13
15	25	20	16
16	30	24	20
17	35	28	23
18	40	33	28
19	46	38	32
20	52	43	38
21	59	49	43
22	66	56	49
23	73	62	55
24	81	69	61
25	89	77	68

From Siegal (1956); after Wilcoxon (1949).

APPENDIX 8

The Kruskal-Wallis Analysis of Variance

Probabilities (p) associated with calculated values of H for three samples from $n_1 = 2, n_2 = 1, n_3 = 1$ to $n_1 = 5, n_2 = 5, n_3 = 5$. The *larger* the value of H the smaller the probability that the null hypothesis is correct.

n_1	n_2	n_3	H	p	n_1	n_2	n_3	H	p
2	1	1	2·7000	0·500	4	1	1	3·5714	0·200
2	2	1	3·6000	0·200	4	2	1	4·8214	0·057
								4·5000	0·076
2	2	2	4·5714	0·067				4·0179	0·114
			3·7143	0·200					
					4	2	2	6·0000	0·014
3	1	1	3·2000	0·300				5·3333	0·033
								5·1250	0·052
3	2	1	4·2857	0·100				4·4583	0·100
			3·8571	0·133				4·1667	0·105
3	2	2	5·3572	0·029	4	3	1	5·8333	0·021
			4·7143	0·048				5·2083	0·050
			4·5000	0·067				5·0000	0·057
			4·4643	0·105				4·0556	0·093
								3·8889	0·129
3	3	1	5·1429	0·043					
			4·5714	0·100	4	3	2	6·4444	0·008
			4·0000	0·129				6·3000	0·011
								5·4444	0·046
3	3	2	6·2500	0·011				5·4000	0·051
			5·3611	0·032				4·5111	0·098
			5·1389	0·061				4·4444	0·102
			4·5556	0·100					
			4·2500	0·121	4	3	3	6·7455	0·010
								6·7091	0·013
3	3	3	7·2000	0·004				5·7909	0·046
			6·4889	0·011				5·7273	0·050
			5·6889	0·029				4·7091	0·092
			5·6000	0·050				4·7000	0·101
			5·0667	0·086					
			4·6222	0·100					

Sample sizes			H	p	Sample sizes			H	p
n_1	n_2	n_3			n_1	n_2	n_3		
4	4	1	6·6667	0·010	5	3	1	6·4000	0·012
			6·1667	0·022				4·9600	0·048
			4·9667	0·048				4·8711	0·052
			4·8667	0·054				4·0178	0·095
			4·1667	0·082				3·8400	0·123
			4·0667	0·102					
					5	3	2	6·9091	0·009
4	4	2	7·0364	0·006				6·8218	0·010
			6·8727	0·011				5·2509	0·049
			5·4545	0·046				5·1055	0·052
			5·2364	0·052				4·6509	0·091
			4·5545	0·098				4·4945	0·101
			4·4455	0·103					
					5	3	3	7·0788	0·009
4	4	3	7·1439	0·010				6·9818	0·011
			7·1364	0·011				5·6485	0·049
			5·5985	0·049				5·5152	0·051
			5·5758	0·051				4·5333	0·097
			4·5455	0·099				4·4121	0·109
			4·4773	0·102					
					5	4	1	6·9545	0·008
4	4	4	7·6538	0·008				6·8400	0·011
			7·5385	0·011				4·9855	0·044
			5·6923	0·049				4·8600	0·056
			5·6538	0·054				3·9873	0·098
			4·6539	0·097				3·9600	0·102
			4·5001	0·104					
					5	4	2	7·2045	0·009
5	1	1	3·8571	0·143				7·1182	0·010
								5·2727	0·049
5	2	1	5·2500	0·036				5·2682	0·050
			5·0000	0·048				4·5409	0·098
			4·4500	0·071				4·5182	0·101
			4·2000	0·095					
			4·0500	0·119	5	4	3	7·4449	0·101
								7·3949	0·011
5	2	2	6·5333	0·008				5·6564	0·049
			6·1333	0·013				5·6308	0·050
			5·1600	0·034				4·5487	0·099
			5·0400	0·056				4·5231	0·103
			4·3733	0·090					
			4·2933	0·122					

Sample size			H	p	Sample size			H	p
n_1	n_2	n_3			n_1	n_2	n_3		
5	4	4	7·7604	0·009	5	5	3	7·5780	0·010
			7·7440	0·011				7·5429	0·010
			5·6571	0·049				5·7055	0·046
			5·6176	0·050				5·6264	0·051
			4·6187	0·100				4·5451	0·100
			4·5527	0·102				4·5363	0·102
5	5	1	7·3091	0·009	5	5	4	7·8229	0·010
			6·8364	0·011				7·7914	0·010
			5·1273	0·046				5·6657	0·049
			4·9091	0·053				5·6429	0·050
			4·1091	0·086				4·5229	0·099
			4·0364	0·105				4·5200	0·101
5	5	2	7·3385	0·010	5	5	5	8·0000	0·009
			7·2692	0·010				7·9800	0·010
			5·3385	0·047				5·7800	0·049
			5·2462	0·051				5·6600	0·051
			4·6231	0·097				4·5600	0·100
			4·5077	0·100				4·5000	0·102

From Siegel (1956); after Kruskal and Wallis (1952).

APPENDIX 9

Spearman's Rank Correlation Coefficient

The critical values of r_s for $N = 4$ to $N = 30$ at the 0·05 and 0·01 levels of significance. The larger the value of r_s the more significant is the result.

For numbers of pairs greater than $N = 30$ the critical value of r_s alters only slightly.

N	Significance level (one-tailed test)	
	0·05	0·01
4	1·000	
5	0·900	1·000
6	0·829	0·943
7	0·714	0·893
8	0·643	0·833
9	0·600	0·783
10	0·564	0·746
12	0·506	0·712
14	0·456	0·645
16	0·425	0·601
18	0·399	0·564
20	0·377	0·534
22	0·359	0·508
24	0·343	0·485
26	0·329	0·465
28	0·317	0·448
30	0·306	0·432

From Siegel (1956); after Olds (1938 and 1949).

APPENDIX 10

Kendall's Rank Correlation Coefficient

Probabilities associated with calculated values of S for $N = 4$ to $N = 10$. The *larger* the value of S for any given value of N the *smaller* the probability that the association of values tested is due to chance or random variations.

S	Values of N 4	5	8	9	S	Values of N 6	7	10
0	0·625	0·592	0·548	0·540	1	0·500	0·500	0·500
2	0·375	0·408	0·452	0·460	3	0·360	0·386	0·431
4	0·167	0·242	0·360	0·381	5	0·235	0·281	0·364
6	0·042	0·117	0·274	0·306	7	0·136	0·191	0·300
8		0·042	0·199	0·238	9	0·068	0·119	0·242
10		0·0083	0·138	0·179	11	0·028	0·068	0·190
12			0·089	0·130	13	0·0083	0·035	0·146
14			0·054	0·090	15	0·0014	0·015	0·108
16			0·031	0·060	17		0·0054	0·078
18			0·016	0·038	19		0·0014	0·054
20			0·0071	0·022	21		0·00020	0·036
22			0·0028	0·012	23			0·023
24			0·00087	0·0063	25			0·014
26			0·00019	0·0029	27			0·0083
28			0·000025	0·0012	29			0·0046
					31			0·0023
					33			0·0011

From Siegal (1956); after Kendall (1948).

APPENDIX 11

The Runs Test

Critical values of r in the runs test for n_1 or n_2 from 2 to 20. Any calculated value of r equal to or smaller than the values given in Table A, or equal to or larger than those given in Table B, is significant at the 0·05 level.

A

n_1 \ n_2	2	3	4	5	6	7	8	9	10	11	12	13	14	15	16	17	18	19	20
2											2	2	2	2	2	2	2	2	2
3					2	2	2	2	2	2	2	2	2	3	3	3	3	3	3
4				2	2	2	3	3	3	3	3	3	3	3	4	4	4	4	4
5			2	2	3	3	3	3	3	4	4	4	4	4	4	4	5	5	5
6		2	2	3	3	3	3	4	4	4	4	5	5	5	5	5	5	6	6
7		2	2	3	3	4	4	4	5	5	5	5	5	6	6	6	6	6	6
8		2	3	3	3	4	4	5	5	5	6	6	6	6	6	7	7	7	7
9		2	3	3	4	4	5	5	5	6	6	6	7	7	7	7	8	8	8
10		2	3	3	4	5	5	5	6	6	7	7	7	7	8	8	8	8	9
11		2	3	4	4	5	5	6	6	7	7	7	8	8	8	9	9	9	9
12	2	2	3	4	4	5	6	6	7	7	7	8	8	8	9	9	9	10	10
13	2	2	3	4	5	5	6	6	7	7	8	8	9	9	9	10	10	10	10
14	2	2	3	4	5	5	6	7	7	8	8	9	9	9	10	10	10	11	11
15	2	3	3	4	5	6	6	7	7	8	8	9	9	10	10	11	11	11	12
16	2	3	4	4	5	6	6	7	8	8	9	9	10	10	11	11	11	12	12
17	2	3	4	4	5	6	7	7	8	9	9	10	10	11	11	11	12	12	13
18	2	3	4	5	5	6	7	8	8	9	9	10	10	11	11	12	12	13	13
19	2	3	4	5	6	6	7	8	8	9	10	10	11	11	12	12	13	13	13
20	2	3	4	5	6	6	7	8	9	9	10	10	11	12	12	13	13	13	14

B

n_1 \ n_2	2	3	4	5	6	7	8	9	10	11	12	13	14	15	16	17	18	19	20
2																			
3																			
4			9	9															
5		9	10	10	11	11													
6		9	10	11	12	12	13	13	13	13									
7		11	12	13	13	14	14	14	14	15	15	15							
8		11	12	13	14	14	15	15	16	16	16	16	17	17	17	17	17		
9					13	14	14	15	16	16	16	17	17	18	18	18	18	18	18
10					13	14	15	16	16	17	17	18	18	18	19	19	19	20	20
11					13	14	15	16	17	17	18	19	19	19	20	20	20	21	21
12					13	14	16	16	17	18	19	19	20	20	21	21	21	22	22
13						15	16	17	18	19	19	20	20	21	21	22	22	23	23
14						15	16	17	18	19	20	20	21	22	22	23	23	23	24
15						15	16	18	18	19	20	21	22	22	23	23	24	24	25
16							17	18	19	20	21	22	22	23	23	24	25	25	25
17							17	18	19	20	21	22	23	23	24	25	25	26	26
18							17	18	19	20	21	22	23	24	25	25	26	26	27
19							17	18	20	21	22	23	23	24	25	26	26	27	27
20							17	18	20	21	22	23	24	25	25	26	27	27	28

From Siegel (1956); after Swed and Eisenhart (1943).

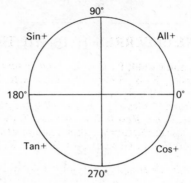

Appendix 12a. Between 270° and 90° the angular measurement is taken from 0°, and between 90° and 270° from 180°.

Values for Sin, Cos, and Tan are positive in the quadrants so marked; for all other values they are negative.

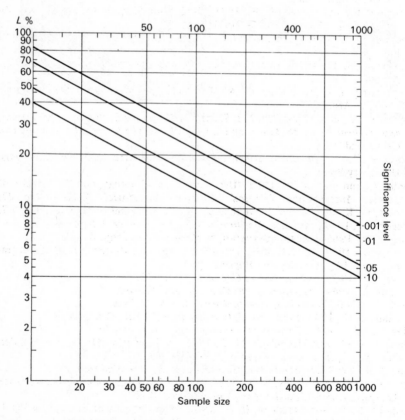

Appendix 12b. Significance of preferred orientation (adapted from Lord Rayleight, 1894, *The Theory of Sound*, and Curray, *Journal of Geology*, 1956, p.126).

WORKS REFERRED TO IN THE TEXT

Abler, R., Adams, J. S., and Gould, P., 1971. *Spatial Organization*. Prentice Hall.

Alder, H. L., and Roessler, E. B., 1964. *Introduction to Probability and Statistics*. W. H. Freeman.

Auble, D., 1953. 'Extended Tables for the Mann-Whitney Statistic'. *Bulletin of the Institute of Educational Research at Indiana University*, Vol. 1, No. 2.

Berry, B. J. L., 1972. 'Cities as Systems within Systems of Cities', reproduced in *The Conceptual Revolution in Geography*, edited W. K. D. Davies, University of London Press.

Carter, H., 1972. *The Study of Urban Geography*. Arnold.

Census of Population 1961. Industry Tables. H.M.S.O.

Census of Population 1971. Preliminary Report. H.M.S.O.

Chorley, R. J., and Haggett, P. (eds.), 1967. *Models in Geography*. Methuen.

Chorley, R. J., and Kennedy, B. A., 1971. *Physical Geography: a Systems Approach*. Prentice Hall.

Clements, F., and Sturgis, R. B., 1971. 'Population Size and Industrial Diversification'. *Urban Studies*, Vol. 8, No. 1, Feb. 1971.

Coppock, J. T., 1964. *An Agricultural Atlas of England and Wales*. Faber and Faber.

Dickinson, G. C., 1970. *Statistical Mapping and the Presentation of Statistics*. Arnold.

Doornkamp, J. C., and King, C. A. M., 1971. *Numerical Analysis in Geomorphology*. Arnold.

Fisher, R. A., and Yates, F., 1953. *Statistical Tables for Biological, Agricultural, and Medical Research*. Oliver and Boyd. 6th ed. 1963 reprinted by Longman Group Ltd. 1973.

Florence, P. S., 1948. *Investment, Location, and Size of Plant*. Cambridge University Press.

Gibbs, J., and Martin, W., 1962. 'Urbanization, Technology, and the Division of Labour: International Patterns'. *American Sociological Review*, Vol. 27.

Gilbert, G. K. 1909. 'The Convexity of Hilltops'. *Journal of Geology*, Vol. 17.

Gregory, S., 1963. *Statistical Methods and the Geographer*. Longmans.

Haggett, P., 1965. *Locational Analysis in Human Geography*. Arnold.

Harvey, D., 1967. 'Models of the evolution of spatial patterns in Human Geography'—ch. 14 in Chorley and Haggett 1967.

Harvey, D., 1969. *Explanation in Geography*. Arnold.

Holmes, A., 1944. *Principles of Physical Geology*. Nelson.

Isard, W., 1960. *Methods of Regional Analysis*. M.I.T. Press.

Kendall, M. G., 1948. *Rank Correlation Methods*. Charles Griffin & Co.

King, C. A. M., 1966. *Techniques in Geomorphology*. Arnold.

King, C. A. M., and McCullagh, M. J., 1971. 'A Simulation Model of a Complex Recurved Spit'. *Journal of Geology*, Vol. 79, No. 1.

King, L. J., 1969. *Statistical Analysis in Geography*. Prentice Hall.

Kruskal, W. H., and Wallis, W. A., 1952. 'Use of Ranks in One-criterion Variance Analysis'. *J. Amer. Statist. Ass.*, Vol. 47, pp.614–17.

Leopold, L. B., and Langbein, H. B., 1962. 'The Concept of Entropy in Landscape Evolution'. *U.S. Geological Survey*. Professional Paper, 500–A.

Lindley, D. V., and Miller, J. C. P., 1953. *Cambridge Elementary Statistical Tables*. Cambridge University Press.

Mann, H. B., and Whitney, D. R., 1947. 'On a test of whether one of two random variables is stochastically larger than the other'. *Ann. Math. Statist.*, Vol. 18, pp.52−4.

Martin, J. E., 1966. *Greater London: An Economic Geography*. Bell.

Merriam, D. F. (ed.), 1970. *Spitsym, a Fortran IV computer program for spit simulation*. Computer Contribution 50. University of Kansas.

Merrington, M., and Thompson, C. M., 1943. Biometrika, Vol. 33.

Moser, C. A., and Scott, W., 1961. *British Towns*. Oliver and Boyd.

O'dell, P., 1974. *Energy: Needs and Resources*. Macmillan.

Olds, E. G., 1938. 'Distributions of Sums of Squares of Rank Differences for Small Numbers of Individuals'. *Ann. Math. Statist.* Vol. 9, pp.133−48.

Olds, E. G., 1949. 'The 5% Significance Levels for Sums of Squares of Rank Differences and a Correction'. *Ann. Math. Statist.*, Vol. 20, pp.117−18.

Pinder, D. A., and Witherick, M. E., 1973. 'Nearest Neighbour Analysis of Linear Point Patterns'. *Tijdschrift voor Economische an Sociale Geographie*, Vol. 64.

Pinder, D. A., and Witherick, M. E., 1975. 'A Modification of Nearest Neighbour Analysis for Use in Linear Situtions'. *Geography*, No. 266, Vol. 60, Part 1.

Readers' Digest 'Book of the Road'.

Robinson, R. J., 1970. *Latin America's Economic Situation. The Use of the Rank Correlation Coefficient*. The Geographical Association.

Siegel, S., 1956. *Nonparametric Statistics for the Behavioural Sciences*. McGraw-Hill.

Smith, D. M., 1969. *Industrial Britain: The North West*. David and Charles.

Swed, Frieda S., and Eisenhart, C., 1943. 'Tables for testing Randomness of Grouping in a sequence of Alternatives'. *Ann. Math. Statist.* Vol. 14, pp.83−6.

Thomas, D., 1963. *Agriculture in Wales during the Napoleonic Wars*. Wales University Press.

Thornthwaite, C. W., 1933. 'The Climate of the Earth'. *Geographical Review*, Vol. 23.

United Nations Statistical Yearbook. 1966.

Weaver, J., 1954. 'Crop Combination Regions in the Middle West'. *Geog. Review*.

Wilcoxon, F., 1949. *Some Rapid Approximate Statistical Procedures*. American Cynamid Co.

World Bank Atlas, 1970. Pub. by the International Bank for Reconstruction and Development.

Young, A., 1963. 'Deductive Models of Slope Evolution'. *3rd Report of the Commission on Slope Evolution of the International Geographical Union: Neue Beiträge zur internationalen Hangforschung*. Vandenhoeck and Ruprecht.

NOTES

CHAPTER 1

1. In practice this distinction is sometimes ignored.

2. It should be noted that Formulas 1.3, and 1.4a, and 1.4b may be used to obtain the standard deviation of any given set of values, whether the values are the whole population, or a sample of the whole population. There is a modified formula, designed to give the 'best estimate' of the standard deviation of the whole population from a sample (normally written $\hat{\sigma}$) which is

$$\hat{\sigma} = \sqrt{\left\{ \frac{\Sigma(x - \bar{x})^2}{n - 1} \right\}}$$

The problems involved in the use of samples are dealt with in detail in Chapter Five. It is necessary to mention this particular point here because some electronic calculators have certain 'automatic' statistical programmes incorporated. When these include the standard deviation, and if the above formula is used, it will give a slightly larger value than Formulas 1.3 or 1.4. Reference should be made to the handbook accompanying the calculator or programme cards.

3. Frequencies in terms of tonnage are not satisfactory as sometimes only the number of items is given, e.g. the *number* of tractors exported is recorded, which cannot be compared with tonnage in other categories.

CHAPTER 2

1. Whether a place is represented by a point or an area on a map depends on the scale of the map. A town may be represented by a single dot, and form part of a point-distribution map of Britain; on a large-scale map of the town, the wards may be the distinct areal units of a choropleth map.

2. This curve was plotted as follows:

Circle, rad. 2 miles contains 80 per cent

∴ Circle, rad. 1 mile contains 20 per cent (80 divided by two squared)

∴ Circle, rad. $\frac{1}{2}$ mile contains 5 per cent (20 divided by two squared)

∴ Circle, rad. $1\frac{1}{2}$ miles, contains 45 per cent (5 times three squared).

3. These terms are also not as straightforward as they seem. The word 'random' is discussed more fully in Chapter Five, but can here be taken to mean 'the result of chance'. The word 'uniform', when applied to a distribution, is here taken to mean: (a) that each individual is the same distance from its nearest neighbour as every other individual; (b) that this nearest neighbour distance is maximized. It can be shown that such a distribution will be triangular (the points will occupy the apexes of equilateral triangles) rather than square. The word 'concentrated' denotes the grouping together of individuals into a single cluster.

4. R in this case is equivalent to the location quotient of the region for the industry in question—i.e. the degree to which regional employment in that industry is above or below the national average. Where the I.Q. is more than 1, regional employment in that industry is above the national average; where less than 1, it is below.

CHAPTER 3

1. If log paper is not available, the data can be plotted logarithmically (using ordinary log. tables) by simply plotting their logarithms on ordinary graph paper. e.g. 10, 100, 1000 are plotted as 1·0, 1·0 and 3·0, so that a constant proportional interval is turned into a constant numerical interval.

CHAPTER 5

1. The rationale for this approximation is as follows: Let p and q be the proportions of textile and non-textile workers in the *population* of workers respectively; let the total sample of workers be n, and the number of textile workers *counted* by T.

Then, the sampling distribution of T will have a mean of np and a standard error of $\sqrt{(npq)}$. The latter can be written as $\sqrt{(np)} \times \sqrt{q}$.

Now, if p is less than $\frac{1}{4}$, q is greater than $\frac{3}{4}$, and $\sqrt{q}$ will approach 1.

Therefore the standard error approximates to $\sqrt{(np)}$. Also, $T = np$. Therefore the standard error approximates to $\sqrt{T}$.

In fact, this is an overestimate, because $\sqrt{q}$ has been taken as 1, and it is actually less than 1. To compensate for this overestimate, the sampling fraction correction factor, which, for a 10 per cent sample should be $\sqrt{0.9}$ (= 0.95), is made to be 0.9, thereby reducing the overestimate. The method provides a short-cut approximation for the standard error. It is worth comparing the result it gives with that derived from the full formula—viz: $\sqrt{(npq)} \times \sqrt{(1-f)}$, which can be written as $\sqrt{npq(1-f)}$. In this example,

$n = 10\,981$ (no. of males in employment in Nottingham counted in census)

$$p = \frac{903}{10\,981} = 0.08 \text{ (proportion of male textile workers)}$$

$$q = (1-p) = 0.92 \quad (1-f) = 0.9$$

Therefore, the standard error = $\sqrt{(10\,981 \times 0.08 \times 0.92 \times 0.9)} = \sqrt{727} \approx 27$.

The result is almost identical to that achieved by the short-cut method.

CHAPTER 6

1. The word 'chance' is probably overused today. It often stands for anything we do not understand or cannot identify. Nevertheless, it is useful shorthand for those factors affecting something we are trying to explain, which are unconnected with the factors we are able to identify and isolate, and which tend to operate in a random fashion upon our object of study. More briefly, 'chance' often means 'extraneous random factors'. In subsequent pages, where the word is used with this connotation, it is put in inverted commas.

2. 'Under H_0' means 'if the null hypothesis were true'.

3. An example of paired data is the traffic flow passing each of a set of census points on two dates, so that each census point is represented by two figures. An example of unpaired data is the traffic flow passing each of two sets of census points on the same date, so that each census point is represented by only one figure.

4. This is a modification of Bessel's correction (see p.150).

5. A two-tailed test is appropriate if we have no reason to predict which of ΣR_1 and ΣR_2 is the smaller and therefore called T (in other words, our alternative hypothesis is non-directional). We can use a one-tailed test if we *have* reason, external to the data, to be able to formulate a directional alternative hypothesis.

CHAPTER 7

1. The product moment coefficient may also be used in partial correlation, using the same formula, substituting r for τ.

CHAPTER 8

1. The reason why the line sought is that which minimizes the sum of the *squares* of residuals is that by so doing a unique and exact solution can be obtained. There may be many solutions if one simply seeks to minimize the sum of residuals. (This is fully discussed in Abler, Adams, and Gould (1971), pp.121–6.)

CHAPTER 10

1.   (i) The vertical loss or gain due to the transport of soil over one section is (in this case) proportional to sin 35°.

(ii) Therefore if k is a constant the vertical gain or loss in one arbitrary time period may be assumed equal to $k \sin 35° t$.

(iii) It follows that if, for convenience, $k = 10$, then units of vertical loss or gain are equal to $10 \sin 35° t$. The units being of equal magnitude to those in which other distances in the model are measured.

ANSWERS TO CONSOLIDATION EXERCISES

Note: Answers are only given where these can be expressed as numbers or short statements.

Chapter 1

1.   *b.* about 50 in 1861
       about 70 in 1971

    c. 23 in 1861
       35 in 1971

    d. uneven.

2.   23·5,   14 . . .

4.   14·8,   20·48% and 12·01%

5.   0·37

6.   Clipstone   4-crop
    Willoughby   1-crop
    Ruddington 3-crop

7.

	Ranked				Scaled				Standardized			
Nigeria	10	9	11	30	2	14	0	16	−1·26	−0·82	−0·83	−2·91
W. Germany	4·5	4	3	11·5	91	81	45	217	+ ·92	+ ·82	+0·67	+2·41
U.K.	2	2·5	2	6·5	96	93	48	237	+1·04	+1·20	+0·76	+3·00
Mexico	6	5	6	17	57	56	9	122	+0·08	+0·08	−0·51	−0·35

Ranks of composite scores:

	From ranks	From scaled scores	From standardized scores
India	8	10	10
U.S.S.R.	3	3	3
U.S.A.	1	1	1
Indonesia	10	11	11
Pakistan	9	8	8
Japan	5	5	5
Brazil	7	7	7
Nigeria	11	9	9
W. Germany	4	4	4
U.K.	2	2	2
Mexico	6	6	6

8. | Nigeria | 29 | 1·46 | 0 |
    | W. Germany | 4850 | 3·69 | 87 |
    | U.K. | 5139 | 3·71 | 88 |
    | Mexico | 1044 | 3·02 | 61 |

8 and 9: Composite scores:

Col. A: from scaled values, incorporating scaled logs of energy consumption (Qu. 8)

Col. B: from weighted ranks (Qu. 9)

	A	B
India	35	51
U.S.S.R.	268	22·5
U.S.A.	295	12
Indonesia	25	56
Pakistan	53	58
Japan	222	22
Brazil	138	40
Nigeria	16	59
W. Germany	259	24·5
U.K.	277	13
Mexico	174	34

Chapter 2

1. Crawley ·018,   Bracknell ·027,   Stevenage ·019

2. 1·55

3.

	Total Mileage	Weighted Mileage
London	950	1730
Birmingham	590	1610
Manchester	470	1930
Liverpool	550	2120
Leeds	500	2200
Newcastle	840	3540

4.

	H	CB	F	C
H	–	125	173	139
CB		–	136	139
F			–	127
C				–

Welsh towns 140
E. Anglian towns 124

(Chapter 2—contd.)

5.

	SL	N	E	L	O
SL	–	0	0	0	0
N		–	1	0	1
E			–	0	1
L				–	0
O					–

C.I = 0·3

6. (a) 1880: 0·36   1940: 0·17

7. (b) Agricultural Employment more evenly distributed.

8. (a) Agriculture: 0·34 ⎫
        Prof. and Scientific services 0·07 ⎰ related to All Employment

   (b) Agriculture: 0·37 ⎫
        All Employment: 0·53 ⎰ related to area

9. 1881: 62·49   1921: 56·53

Chapter 3

1.

	Ruddington	Nottingham
1901	81·4	86·8
1911	90·4	94·1
1921	93·9	95·1
1931	100·0	100·0
1951	147·9	110·8
1961	169·2	112·9

4. 1916–27 $y = 82·9x + 2313$
    1928–39 $y = 116·3x + 3195$

5. *Linear:* $y = 2·8x + 23·4$
        1911 . . . 34·6 mill
        1921 . . . 37·4 mill

   Logarithmic: $\log y = 0·0525x + 1·357$
        1911 . . . 36·7 mill (antilog. of 1·567)
        1921 . . . 41·6 mill (antilog. of 1·619)

6.

	2000 A.D.	2030 A.D.
2% p.a.	6308 m.	11420 m.
1·8% p.a.	6032 m.	10300 m.
1·6% p.a.	5764 m.	9276 m.

(Chapter 3—contd.)

7. England and Wales   1·2%
    Nottingham   2·1%
    Ruddington   0·6%

8. Total energy   131·5   times
    Population   7·24 times
    p.c. energy   18·15 times

Chapter 4

1. (a) 100   (b) 1/6 or 0·17

2. (a) 106/206   (b) 100/206   (c) 206/206 = 1

3. 1/6

4. (a) 1/6   (b) 1/12   (c) 5/18

5. q^4, $4q^3p$, $6q^2p^2$, $4qp^3$, p^4

7. $a = 0·025$ $b = 0·135$ $c = 0·34$

8. (i) 0·025   (ii) 0·16   (iii) 0·5   (iv) 0·84   (v) 0·975

9. (i) 0·975   (ii) 0·84   (iii) 0·5   (iv) 0·16   (v) 0·025

10. (i) 0·16   (ii) 0·025   (iii) 0·34   (iv) 0   (v) 0·475

11. (i) 0·27   (ii) 0·75

12. (a) (i)   1·2% or 0·012
        (ii)   6% or 0·06
        (iii)   47% or 0·47
        (iv)   38% or 0·38
    (b) (i)   0·22   (ii) 0·0034

13. Minimum: 19·6 inches
    Maximum: 31·0 inches

Chapter 5

2. (i) Pop: All pebbles on beach
        S.F.:   All pebbles touching cross located at all points on beach
    (ii) Pop: All buildings in city
        S.F.:   Buildings at all combinations of distance and direction from city centre

(Chapter 5—contd.)

    (iii) Pop: All locations on mapped area
           S.F.: All sections across map

3.   1·41   47·2 and 52·8   5·6
      1·16   47·7 and 53·3   4·6
      1·00   48·0 and 52·0   4·0

4.   (a) 1950 and 2050
    (b) 4 times

5.   a   40 ± 2·75 (small sample method)—correct
          40 ± 2·50 (large sample method)—wrong
    b   40 ± 2·10 (small sample method)—correct
          40 ± 2·00 (large sample method)—wrong

% Error of large sample method: 9·1 in a, 4·8 in b.

6.

n	p	q	Mean %	S.E.%	95% CL	CI
100	·8	·2	80	4	80 ± 8	16
100	·6	·4	60	4·9	60 ± 9·8	19·6
100	·5	·5	50	5	50 ± 10	20
100	·4	·6	40	4·9	40 ± 9·8	19·6
100	·2	·8	20	4	20 ± 8·0	16

CI inversely proportional to square root of sample size and increases as p approaches ·5 from above and below.

Chapter 6

1. $p > 0·10$ Not significant

2.   Significant at ·001 level

3.   (i) Arable: significant ($z = 2·04$)
    (ii) Grass: not significant ($z = 1·06$)
    (iii) Wood and Rough Pasture: Comparison not possible npq < 9

4.   A and B not significant ($p = ·134$—two-tailed)
    B and C not significant ($p = ·368$—two-tailed)
    A and C significant at ·016 level (two-tailed)

5.   (a) Significant at ·016 level (two-tailed)
    (b) Significant at ·028 level (two-tailed)

6.   Significant at ·019 level (two-tailed)

(*Chapter 6—contd.*)

7. Significant at ·05 level (two-tailed), but not at ·02 (result depends on sample taken)

8. Significant at ·05 level (two-tailed), but not at ·02

9. Significant at ·05 level (two-tailed)

Chapter 7

1. $r = -0.47$

2. Using r_s, protein and population growth significant at ·05 level; calories and population growth is not

3. Significant at ·01 level

4. τ for sinuosity and sand (wetted perimeter const.) $= \cdot 26$
   τ for wetted perimeter and sand (sinuosity const.) $= \cdot 5$

5. $r = \cdot 58$

   H_0 is that sample values from north and south facing slopes are drawn from the same population, and that differences are due to random variations

   H_1 is that differences in the two data sets are significant

6. ·05 level

7. t-test for independent samples
   Mann Whitney U-test
   Chi Square Test
   Binomial Test

Chapter 8

3. $y = 1 \cdot 08x + 1 \cdot 71$
   $x = 0 \cdot 76y + 0 \cdot 04$

4. (a) 10·11
   (b) Hastings and Colchester

Chapter 9

1. $R = 1 \cdot 9 \ldots$
   More significant than ·001 level

3. 95%

INDEX